SCIENCE, GOD AND THE
NATURE OF REALITY

SCIENCE, GOD AND THE NATURE OF REALITY

BIAS IN BIOMEDICAL RESEARCH

SARAH S. KNOX

BrownWalker Press
Boca Raton

Science, God and the Nature of Reality:
Bias in Biomedical Research

BrownWalker Press
Boca Raton, Florida • USA
2010

ISBN-10: 1-59942-545-9 *(paper)*
ISBN-13: 978-1-59942-545-0 *(paper)*

ISBN-10: 1-59942-546-7 *(ebook)*
ISBN-13: 978-1-59942-546-7 *(ebook)*

www.brownwalker.com

Publisher's Cataloging-in-Publication Data

Knox, Sarah S.
Science, God and the nature of reality : bias in biomedical research / Sarah S. Knox.
p. cm.
ISBN: 1-59942-545-9
1. Religion and science. 2. Evolution (Biology)—Religious aspects. I. Title.
BL240.3 .K61 2010
213—dc22

2010936951

To Eje, Mary, Judy, Robin and Eloise

◆

May their suffering not have been in vain

CONTENTS

♦

INTRODUCTION
◆
PREMISE AND HYPOTHESES

"...everyone who is seriously involved in the pursuit of science becomes convinced that a spirit is manifest in the laws of the universe, a spirit that is enormously superior to that of human beings, in the face of whom we must reflect humbly on our modest powers."

—Albert Einstein [I.1]

The purpose of this book is to begin a scientific dialogue. The subject of the dialogue is the existence or non-existence of God as a legitimate focus of scientific inquiry. The book does not purport to prove the author's opinion concerning God's existence, but rather to clarify the fact that the prevailing assumption in scientific circles that God does not exist or is irrelevant to the pursuit of scientific knowledge, is a cultural belief not a fact supported by scientific data. However, this belief is so pervasive that it has become axiomatic in every field of scientific research, sculpting the nature of the questions scientists ask and the variables they include as potential explanations in scientific investigations. It is therefore, a potent and unresolved issue.

Let us examine why. A current axiom in the field of medicine is that living organisms originated from inorganic matter that, through a series of undetermined circumstances (e.g., temperature, chemical mixtures, surrounding environment, electrical storms, etc.) randomly evolved into a living cell which became the source of all life. Through some as yet undefined process, this cell gained the ability to reproduce itself and through random mutations evolved through many plant and animal stages to become human beings. Thus, the source (cause) of not only our body, but our thoughts, feelings and cognitions is particulate matter. It is assumed that the organs and systems in our bodies are made up of small particles of matter and that the most logical way to study the origins of disease is to begin with the smallest possible unit and extrapolate from there to the diseased system. It is assumed that if one can understand what has gone wrong with the subunits such, as a cells, receptors, neurotransmitters, DNA, etc., one will be able to understand the cause and treatment of the disease itself. Given the premise that the whole is no more than a sum of its individual parts, this is certainly a logical approach. Referred to as "reductionism," it is also the most common investigative approach in biomedical research. If one believes that understanding a complex system can best be achieved by studying its smallest individual components, it would not make sense to try to understand a diseased organ by beginning with an examination of systemic functioning in anything other than the diseased organ, and certainly not by inquiring about the ill person's recent life events or emotional state because the cause of the problem is assumed to originate locally, in some physical part of the organ that is manifesting the disease. Causality is seen as unidirectional, emanating from a localized malfunction, such as a gene, receptor, signaling pathway, mutated cell, or diseased organ but having the potential for spreading to other parts of the body. The possibility that imbalance in the integration of multiple complex systems could be the cause that allows what would otherwise be an easily repaired malfunction to progress to the stage of disease (such as a tumor) in a single organ, is not consistent with the reductionist approach nor with the fundamental tenets of current biomedical research.

If the scientist investigating the disease is not a reductionist, s/he believes that the whole is more than the sum of its parts. This means that the system has properties that cannot be extrapolated from its individual components. Such a scientist might take a more holistic approach to the investigation of disease. Assuming that humans are complex physical, psychological, emotional, (and spiritual?) beings, this scientist might begin by asking what part of the complex system is out of balance when symptoms appear and why the disease is manifesting in that particular part of the body at that particular point in time. The latter approach would only be utilized if one believed (hypothesized) that physical and emotional aspects of "humaness" are so intertwined that examining only the physical aspects of one small part provides insufficient information to fully understand why a particular individual is ill at a specific point in time. Let us take the example of "strep throat". We know that streptococcal bacteria are associated with this illness and can identify them easily under a microscope. What is less clear is why when two healthy people are exposed to these bacteria in the same manner, only one becomes ill. Traditional medicine would say that the immune system was probably compromised in the person who became sick due to lack of sleep or poor diet. Although stress also influences immune function, asking questions about a patient's life situation is almost never part of a routine physical examination. Given the large number of bacilli with which the average person comes in contact on a daily basis, and the relatively healthy state of most individuals, one has to acknowledge, that for the most part, the immune system is a very effective "watch dog". So when someone becomes ill, one can certainly limit treatment to an antibiotic, if it is a disease which responds to antibiotics. But having some insight into why the immune system is not responding as well as it should at a particular point in time, can also help to identify other factors contributing to disease vulnerability and promote preventive measures that will avoid a recurrence in the future.

Research into the influence of emotions and cognitions on physiological functioning has been the purview of Behavioral Medicine for several decades but has never been fully accepted or integrated into the thinking of the biomedical research community [I.2]. I believe that there is an underlying reason for this. The reductionist viewpoint posits the whole as simply the sum of its parts, which in turn, are composed of particulate matter. If causality is unidirectional, i.e., goes from the smallest to the largest, and matter is the only cause, then thoughts and feelings cannot possibly have any meaningful influence on physiological functioning. (The belief in matter as the sole cause of everything is referred to as "materialism" by theorists and differs from the more common use of the word "*materialism,*" which refers to a value system focused on accumulating possessions.) The materialist / reductionist approach to science differs radically from the holistic point of view that is based partially on complexity theory. The holistic approach is that matter not only influences thoughts and feelings *but can be influenced by them.* In the specific case of a streptococcal infection, the holistic approach would consider the fact that thoughts and feelings such as stress can cause changes in the neuroendocrine chemistry of the brain, which results in a cascade of changes influencing immune function. Therefore, the holistic approach would ask not only whether a streptococcal culture is positive, but why the immune defenses are not capable of defending against it at that particular point in time. What implications do these different approaches have for treatment? In the reductionist approach, one would undoubtedly turn first to antibiotics to cure the strep infection. The holistic approach would probably also recommend antibiotics as the first step in treatment but might also attempt to ascertain why the immune system had failed (e.g., diet, stress, passive smoking, lack of sleep, etc.) and how this could be prevented in the future. In the case of recurring childhood illnesses such as earache, the latter approach might provide additional benefit.

The way that beliefs influence scientific inquiry can be further elucidated by contemplating the most common approaches to understanding human thinking and consciousness. The reductionists posit that the brain, neurochemicals and nerves are the cause of thought. They describe in detail blood flow changes in different parts of the brain, the chemical changes involved in the propagation of a nerve impulse: e.g., how potassium and sodium are transported across the membrane, the role of calcium channels and multiple chemical changes involving neurotransmitter substances that occur at nerve endings and in the synapses between nerves. However, there are no data that relate the voltage across the nerve or the amount of a specific neurotransmitter at the end of an axon to a specific thought. No data exist that would indicate that the propagation of a nerve impulse along an axon in the process of thinking about a chocolate cake recipe differs in any way from a nerve impulse created by reflection on a Shakespearean sonnet or a porno movie. In fact, I know of no neurophysiological research that has ever explicitly defined what a thought is.

The fact that neurophysiologists have not even come close to defining thought seems strangely, to be of little concern to them or to any other field of neuroscience or psychiatry involved in describing brain function or treating mental disorders. We know that there are particular areas of the brain associated with visual, auditory, emotional and other processes. But we cannot define thought. We know that a state of emotional distress is associated with changes in the chemistry of certain areas of the brain, such as the locus coeruleus, hypothalamus or pituitary, and that these areas are connected to other parts of the body which can result in a cascade of nervous system and endocrine influences on areas such as the cardiovascular and immune systems. The data also indicate that there is reciprocity in these functions such that changes in brain chemistry can affect the way one feels and that feelings can cause changes in brain chemistry. Thus, negative thoughts and feelings can cause chemical imbalance in a previously healthy system, just as chemical imbalance can cause negative thoughts and feelings. However it is not possible to look at the amount of a particular neuroendocrine secretion and know what the person was thinking or experiencing. Other mammals with similar neuroendocrine secretions do not think like humans, nor do they have comparable intelligence, which leads to the conclusion that these chemical changes might not be a sufficient condition for thinking to occur.

If thoughts and feelings can change chemistry, but we cannot reduce thoughts and feelings to chemistry, then logic suggests that either there must be important, as yet undiscovered chemicals that explain these processes or that maybe we should be seeking the answer elsewhere. One possibility is that patterns of electrical impulses associated with thoughts, that is to say, varying wavelengths of electromagnetic fields (energy) in interaction, might in some way be important to understanding conscious thought. An even more radical possibility is that mind, like matter (as defined by quantum theory), is non-local (i.e., not confined to the brain). These are radical concepts that will be explored more deeply in the remainder of this book.

If we step back for a minute to examine what science is really about, its purpose is to try, to the best of its ability, to explain the nature of reality. The more accurately we can do this, the more problems we will be able to solve and the more our research will benefit humanity. The premise of this book is that there are major differences in the way that modern physics and the biological / medical sciences define the essence of reality and that it is time to step back and take a look at the consequences of this divergence. The lack of knowledge among biomedical scientists concerning what has been experimentally demonstrated in modern physics is creating a bias in the biomedical approach to research. If our scientific questions are biased by our personal beliefs, the process of science and thus, the conclusions that we draw will be flawed.

To further illustrate this, we might ponder how a scientist with a materialist / reductionist viewpoint and one with a holistic view would go about trying to predict how a building would react under conditions of strong wind. Just as a biomedical reductionist might try to understand a particular disease by examining the molecular structure of a diseased organ, the engineer using a reductionist approach might begin by examining whether screws with the appropriate composition, width and length have been chosen for the material in the frame. If these seem to be adequate for the size and strength of the weight they must bear, then the steel beams might be the next level of investigation, and so it would continue through choice of windows, doors, etc. until all the individual building components had been examined. The holistic theorist, on the other hand might begin by looking at the blueprint of the building to see whether the construction design was appropriate for the geographical region and weather conditions in the location intended for construction. Is this an earth quake zone? If so, does the design allow for movement? Does it need to be changed to accommodate periodic flooding? Translated into the realm of biomedical research, the holistic scientist would assume that there is a systemic response when the body is confronted with a disease pathogen, and that the most logical way to solve the riddle of a particular disease would be to understand how the subsystems interact with the whole when the body is exposed. Since the immune system, DNA repair mechanisms and mechanisms of programmed cell death (apoptosis) are facile enough to respond to foreign invaders, we do not usually develop symptoms or become ill as a result of mistakes in DNA replication, renegade malignant cells, or the plethora of bacteria and viruses to which we are exposed in our daily contact with the environment. Nor do we usually get cancer from the mutational "mistakes" made in the continually ongoing process of cell division that replaces aging cells. When the coordination and feedback between various defense mechanisms of the body are working properly, the 'marauder' cells or mutations are sought out and destroyed, and we are none the wiser. This example illustrates that different assumptions about causality can lead to different investigative approaches and to the inclusion of different variables in the experiments.

What should by now be clear to the thoughtful reader is that both of these methods, namely, studying the individual pathogen and investigating the host system, supply us with important and useful information for trying to understand the cause and find the cure for disease. Employing one method to the exclusion of the other, will obscure important data and leave us with incomplete answers and suboptimal treatments. Yet, there is a subtle bias in biomedical research, which has caused the primary emphasis to be placed on the disease-causing agent and on analyzing the response at a molecular or organ specific level, while ignoring the integrated systems that provide the functional context of the body's defense against illness.

Returning to the concept of how our belief systems influence the way we conduct scientific research, most scientists seem to be in agreement that the cosmos and life on earth evolved as a result of random events and that design was not involved. Since there is general agreement on this point, one would assume that there was also abundant scientific data to support it. Indeed, there is abundant data to support the concept of evolution from simple to more complex organisms and data that supply information concerning which species came first and approximately when in the earth's evolutionary history different species emerged. However, to my knowledge there is no data whatsoever that would support a theory of randomness vs. design. Dominant theories concerning the origin of life are still unsubstantiated and without supporting documentation. In other words, they are based purely on speculation. The fact that most scientists adhere to the former belief (randomness) and most religious people to the latter (design), does not change the fact that neither can be proven with scientific evidence. Therefore, although most scientists would insist that they are "objective" and that they approach scientific questions with an open mind, one of the most fundamental assumptions of current scientific theories is founded on conjecture.

For centuries, there was general agreement in European thought that God had created the universe and everything in it. It was the task of the learned to figure out how it worked. However, when data began to contradict the teachings of the Catholic Church, science began to emerge as a school of thought separate and distinct from religion. A major split occurred with the teachings of Copernicus and Galileo, whose views that the earth was not the center of the universe were rejected by the church because they disagreed with its doctrine. Galileo's telescopic observations of the heavens and subsequent calculations concerning celestial movement supported Copernicus's theory that the earth rotated around the sun rather than the sun around the earth. He was condemned by the inquisition for supporting the Copernican view and was silenced for the remainder of his life. Religion and science were no longer compatible.

Newton, too, believed in God, but despite his belief in God's role as creator of the universe, his physics, describing general principles for how energy and matter interacted, seemed to make the existence of God irrelevant. *If* one refrained from asking how these principles came to be in the first place (and one did), one could eliminate God from the equation altogether. The world as Newton knew it functioned according to invariant principles, like a clock, and these principles could be understood on their own merit without invoking intervention from God. The subsequent work of Darwin on the "survival of the fittest" principle of evolution and Mendel (genetic inheritance) cemented materialism as the scientific explanation for the nature of reality.

What has changed since Newton and Darwin is the discovery and experimental validation of quantum mechanics. Quantum theory turned the materialist view of reality on its head by showing that at a subatomic level, distinctions between matter and energy blur. In fact, one of the most well-known quantum physical experiments showed that whether light consisted of particles or waves (matter or energy) depended solely on how the experiment was set up. This completely contradicted the world of Newtonian physics, which defined reality as an objective state, totally independent of the observer. The problem presented by quantum theoretical experiments was that on a subatomic level, the little building blocks of matter disappear, as energy and matter become two aspects of one and the same reality. Furthermore, whether reality appears as particles or waves, depends on how the measurements are made, implying that there is no 'objective reality' apart from the observer. The enormous implications of these principles have gone unrecognized by the field of biomedical research. The consequence is that the primary theoretical framework of biomedicine, namely materialism, is more than 80 years out of date. The purpose of this book is to discuss these developments and to call attention to the gap between the nature of reality as defined by experimental data from modern physics and the belief system that still poses as science in the fields of biology and medicine. The fundamental basis for refusing to acknowledge the scientific relevance of the question of Divine design vs. randomness in evolution is the doctrine of materialism, which is no longer a valid scientific theory. Despite the predominant belief to the contrary, the question of God's existence or non-existence is still unresolved.

So what relevance does this have for science? Our beliefs influence not only the questions we consider worthy of scientific inquiry, but also the methods we use to investigate them. The subatomic blurring of the distinction between matter and energy raises challenges to many of our most cherished assumptions, not the least of which is our tendency to attribute all causality to matter. Einstein long ago showed us that matter and energy are interchangeable. What quantum theory leaves open is the distinct possibility that energy may actually be primary and matter secondary. Does energy "congeal" (for lack of a better word) into matter at a lower vibrational level or are matter and energy always co-existent? The lack of mass in a photon opens the intriguing possibility that energy may be primary. What implications do these issues have for evolution? If quantum theory illustrates one thing, it is that there are no indestructible bits of matter that constitute the basic building blocks of all substance. This being the case, where does causality lie with respect to life? The issue of design in the universe is still unresolved and so is the issue of God's existence. Like many other formerly philosophical problems that made the transition into the scientific realm as we developed the methodology to investigate them, the issue of God's existence is one that should no longer be left to the purview of philosophy and theology. The implications of this question for human existence are simply too momentous to be ignored by science. Whether or not there is a God and the role (if any) that this God plays in the universe is the most important scientific issue in existence because it is fundamental to everything else we are trying to understand.

Beliefs about the origin of the universe, e.g., whether its laws were developed randomly or as part of a complex creative process orchestrated by design, have important implications for how we structure scientific inquiry and the methodology we use to investigate the nature of reality. If the majority of scientists have the same belief system and this belief is incorrect, the process of scientific discovery will be seriously impeded. For this reason, it is important that those of us who are scientists become aware of and openly acknowledge our beliefs and begin to reflect on the manner in which they might influence the framework of our scientific questions.

The scientific term for unproven beliefs about the nature of reality is "hypotheses." Hypotheses are theories that we generate to explain the nature of observed phenomena whose origin or function we do not fully understand. When we have observed phenomena that puzzle us, we reflect upon them in the context of what we already know and what we believe. Both knowledge and belief influence the way we formulate hypotheses to explain the phenomena, and also the way we design experiments to test their validity. In the formulation of hypotheses, we exercise logic to put the new phenomena in a context that makes sense. Scientific bias influences not only scientific method but also the very questions that are deemed relevant to ask. Because the purpose of this book is to help scientists get beyond their biases through the process of acknowledging them and contemplating the influences these biases have on their research, I will begin by stating my own.

My hypothesis is that the fundamental causality of the universe and of the principles determining its function is Mind (Collective Unconscious, God). I also believe that humans have a divine aspect or "soul" which is their essence and which allows them to become co-creators with God. This belief system is important for my approach to scientific investigation because it influences the questions I see as scientifically relevant (e.g., different forms of energy healing). Because a major assumption of biomedical research, i.e., materialism, has been disproved but not discarded, I believe that it is time to examine the consequences that scientific bias has for scientific inquiry.

There is another primary assumption underlying research in the biomedical sciences, namely the assumption that God is either non-existent or irrelevant to our scientific understanding of the nature of reality. Although I can find no data to either prove or disprove this assumption concerning God's existence, it is axiomatic in biomedical research. The frame of reference created by this assumption determines the selection of relevant lines of investigation, methodological approaches and variables that are included in experiments and equations. Just as my beliefs influence the scientific questions I am asking, the materialist belief system of mainstream science influences the questions it asks *and those it chooses not to ask*. The questions not being asked and the consequences of not asking them are the topic of this book.

I have formulated my hypothesis concerning God's existence as a theory to explain my observations. It occurred after many years of agnosticism and lack of interest in the question, when continued reflection on the nature of reality as reflected in the laws of physics, coupled with the biological data I was observing, finally made non-belief too difficult to sustain. I encourage scientists with opposing views to express their own hypotheses (belief systems) along with the data used in their formulation. My expectation is not that we will then "have the answer" or be able to construct the ultimate experiment to test the existence of God, but rather that the process will result in the pursuit of new scientific questions and an increased understanding of how our beliefs are influencing the way we conduct scientific research. The objective is to broaden the scope of the scientific questions we are asking, and hopefully, to increase the benefit of science for humankind.

The remainder of the book will be devoted to a description of my scientific journey in hopes that it will open the question of God's existence, as well as the implications of this question, to legitimate scientific inquiry. I believe that a constructive dialogue on this subject could lead to research that has the potential for greatly benefiting humanity.

REFERENCES

I.1)   Einstein, A. Letter to a Sunday School child in New York City, January 24, 1936. Quoted in Dukes H. and Hoffman B. *Albert Einstein: The Human Side. New Glimpses from his Archives.* (Translation by Sarah Knox) Princeton University Press, 1979.

I.2)   Relman AS, Angell M. Resolved: Psychosocial interventions can improve clinical outcomes in organic disease (CON). *Psychosom Med* 2002; 64: 552-563.

CHAPTER 1

THE NATURE OF REALITY:
IMPLICATIONS FOR SCIENTIFIC INQUIRY

1.1. DEFINITION OF MATTER

> "We can say that all particles are made of the same fundamental substance, which can be designated energy or matter; or we can put things as follows: the basic substance 'energy' becomes 'matter' by assuming the form of an elementary particle."
>
> —Werner Heisenberg [1.1]

We have said that the goal of science is to describe as accurately as possible the nature of reality. The goal of medical research is to utilize its discoveries about the nature of living organisms to benefit humanity through helping to design clinical interventions and technical products that improve the quality and length of human life. Since the objective is to understand the nature of reality, and since our bodies and the things around us are composed of matter, we will begin with matter. According to classical physics, matter consists of little units called atoms, containing a nucleus surrounded by electrons. Modern physics has since demonstrated that there are even smaller particles, and even stranger, that matter can also assume the characteristics of a waveform (energy). In fact, at a subatomic level, whether we observe a particle or a waveform depends entirely on the way we set up the experiment. This phenomenon is called "wave / particle duality". In concrete terms, it means that light appears as particles if you measure it one way but as waves if you measure it another way. Modern physics tells us that this is not due to measurement error but to the inherent nature of reality, which does not assume the form of matter or energy until the moment of measurement. If what we are measuring does not become a particle (matter) or a wave form (energy) until we measure it, then *there is no "objective" reality out there.*

How is this possible? It is completely counterintuitive and implies that the "essence of reality" is totally inconsistent with the world that we have come to know through our five senses. We saw a sofa in our living room yesterday and it is still there today. Is this not proof of an objective reality? The whole concept appears to be nonsense. "Objective" reality (e.g., the sofa) sure seems to be where we left it yesterday. Is this wave/particle duality stuff just philosophical 'gobbledygook' or is it really relevant to our daily lives? The answer is that it has all the relevance in the world for how we are conducting biomedical research and the reasons will unfold as the book progresses.

The discrepancy between what we perceive with our five senses and what experiments have verified to be true on a quantum level, stems from the fact that our senses have evolved to serve as practical tools for navigating the daily tasks of lives lived in a 'macro', rather than a subatomic world. They simply do not have the precision to perceive what happens on a subatomic level without the help of additional measuring instruments. The eyesight that helps us to steer a car while we are driving and to cross the street without being hit, is not sensitive enough to allow us to distinguish very small objects, such as bacteria, without using a microscope. This is probably a good thing because what we need to function effectively is the ability to simultaneously observe large areas of our surrounding environment. Just imagine where we would be without peripheral vision. If our eyesight could zero in on activity at a subatomic level, we would be focusing in the same way that one does with a microscope, i.e., on a very tiny spot. We would lose the ability to take in our surrounding environment and thus, be totally oblivious to that truck coming down the street that we are beginning to cross. Our chances of survival would be greatly reduced. The analogy can be made to standing on a hilltop and looking through a camera lens. If we use the camera without the zoom lens, we see a broad landscape, including the tornado heading our way. If we zoom the camera in on a flower in the distance, it looks very beautiful but we miss the tornado altogether. We need to be able to see the tornado. Our eyes function like the lens without a zoom.

So, although we may know from scientific evidence that the separation between the nucleus of an atom and its surrounding electrons contains a great deal of empty space, we have difficulty conceptualizing it because our vision tells us something different. To us, the chair looks solid. We are accustomed to thinking that there are objective bits of matter "out there" whose existence is independent over time, so that they will be the same whether we measure them today, tomorrow, or next week. However, the Schrödinger equation, which provides the most accurate way of predicting the probable state of a measured particle, says something about the nature of reality at a subatomic level that is difficult to grasp. It says that it is basically a linear superposition of *possible* states. That means that the best prediction for finding an atom at a particular place at a particular point in time is the sum of all the possibilities, each multiplied by its probability. So, theoretically, there is a possibility that an electron that is currently in my body may be orbiting somewhere in the vicinity of the moon within the next two minutes, but this is highly *improbable*, given its current position. The possibility multiplied by its probability (e.g., its current location, the distance to the moon, the speed it would have to travel to get there in the next two minutes) results in a highly unlikely occurrence, and one that would be negligible when summed into Schrödinger's equation. The probability that the same electron will still be somewhere in the vicinity of my body would have a much higher probability and thus have more "weight" in the equation. So, although the probability of an electron being at a particular place at a particular point in time is essentially infinite, it decreases drastically with divergence from immediately preceding measured states. At the time of measurement, the multiple possibilities "collapse" into one state (e.g., particle or waveform), determined by a combination of the point in time when the measurement was made, the method of measurement, etc. At the time of measurement one potential state becomes an actuality with specific parameters, while the others disappear as possibilities.

We influence what we measure because the more precisely we measure location at a specific time, the less accurately we can measure the velocity, and vice versa. To measure the location of an electron requires shining a beam of light on it so we can see it, but in so doing we perturb (influence) what we are trying to measure. The smallest unit of light possible, which is required for the most precise measurement at the subatomic level, is one photon. The wave probability associated with this single photon is short (i.e., high frequency) which means that its power to knock the electron at the time of measurement, and change its velocity is high. So by measuring its position with precision we simultaneously change its velocity, making that measurement less precise. Either we can measure the position at a particular point in time with exactness or the velocity can be measured with exactness, but not both. The more precisely we measure one, the less exact will be our measurement of the other.

It was originally thought by some physicists, including Einstein, that even though the Schrödinger equation is the most accurate way we have of predicting future events at a subatomic level, this way of defining reality was really the result of a measurement problem. These scientists believed that there really was an "objective" reality out there and that if we only had better ways of measuring it, all this "potentiality" stuff would disappear and we could return to the ordered, knowable, Newtonian universe. Einstein's famous quote, that "God does not play dice" expressed this sentiment.

However, the hope that this would ultimately be resolved as a problem of measurement error has been dashed by the experimental evidence [1.2]. More than three quarters of a century have elapsed and the accuracy of this equation has been demonstrated by repeated, rigorous experimental evidence and by a proliferation of technological innovations based on quantum mechanics. Hard as it is to grasp, *multiple potentiality is inherent in the nature of reality*. There is no "objective reality" independent of measurement. What this implies is that the method of measurement is extremely important for the outcome. If we set up the light experiment to measure particles, we will see only particles. If we set it up to measure wave forms, light will appear as wave forms, but not particles. This has important implications for research and particularly basic biomedical research, which is primarily designed to see particles (receptors, DNA base pair sequences, etc.). Given the dual nature of reality, it is important to ask whether or not setting up experiments so that we only see particulate aspects of matter is obscuring other important information that could improve clinical outcomes.

The profound implications of quantum mechanics have been extensively discussed by physicists for many years. One prominent physicist, Hans-Peter Durr [1.3], retired Director of the Werner Heisenberg Max Plank Institute in Germany and co-author of a number of papers with Werner Heisenberg, says that, "Einstein's special theory of relativity, however, enforced another important change by revealing mass as a special form of concentrated energy. As a consequence there is no principal difference between matter and force field anymore..." This statement actually makes energy the primary substance which then "congeals" into matter (see also Heisenberg, above). So rather than matter being the most fundamental substance, as is believed in biomedical research, energy is depicted as the primary focus. Durr says that the wave/particle duality "forced physics out of its old setting as a reality of interacting objects, into a new setting of mere "potentiality" which, under special conditions is capable of coagulating into reality." At the moment of measurement, the multiple possibilities disappear and potentiality becomes reality. Durr draws the conclusion that to the extent that we can measure individual particles, they are excerpted from a context or "whole." Thus, he also turns the tables on reductionism by implying that accurate knowledge of the particle can only be achieved by first understanding the larger context from which it was extracted. What comes first is a state of multiple potentiality. Only by understanding the context of this multiplicity, can we hope to understand the role of the singularity that has been extracted. If we return to our first example of how a building reacts in strong wind, we will not understand the role of the screw until we have seen the blueprint and know how the screw fits into the context of the design. If the screw is appropriate for the design but the design is not appropriate for the geographical region, then examining the screws will not help us.

As the quotes from both Durr and Heisenberg demonstrate, the important implication of quantum physics is that materialism, i.e., the theory that all phenomena in the universe, including mind, have their causality in matter, is not an accurate depiction of reality. Either there is one fundamental substance that can assume the form of either energy or matter, depending on how it is measured, or, as Heisenberg stated, "the basic substance 'energy' becomes 'matter' by assuming the form of an elementary particle" [1.1]. This has profound implications for biology and medical science, both of which assume the opposite, namely that matter is primary. This (incorrect) assumption leads biomedical scientists to automatically assume that the different forms of energy measured in and around the body (e.g., heart rhythms, brain waves, and electrical properties associated with skin conductance and muscle tension) result from or are artifacts of matter, and do not play a causal role in either organ functioning or the disease process. This assumption is so universal in medical research that it has assumed the form of an axiom which no one has actually bothered to verify experimentally. Because physics is not a part of most medical and graduate school curricula, the implications of quantum theory for biology and medicine have been lost. Materialism is still the doctrine that dominates scientific thinking in these fields and dictates the definition of what constitutes plausible mechanisms and good science. Science is not as objective in its methodology as it would have us believe.

The extent to which the materialist belief system influences scientific investigation can be illustrated by an experiment that took place several years ago. A study was conducted in which all patients entering a coronary care unit at San Francisco General Hospital for a period of ten months were randomized to either a treatment or a control group [1.4]. Although the nature of the study was explained to the patients, both they and their doctors were blinded as to which patients were in the experimental group. Both groups received standard care. In addition, the experimental group received intercessory prayer. Each patient was assigned to several intercessors who were asked to pray daily during the patient's hospital stay for a rapid recovery and for prevention of complications and death. They never met the patient. The intercessors were "born again" Christians, who were given the patients' first name, diagnosis, and regular progress reports. They were free to add anything else into the prayers that they chose. The praying was done outside the hospital. Statistical analyses showed no differences in severity of illness between patient groups at the time of entry into the study. By the end of the study, the group that was prayed for had experienced fewer incidents of cardiopulmonary arrest and pneumonia, had taken fewer diuretics and antibiotics, and had required fewer numbers of intubation/ventilation procedures than the control group. These results cannot be written off as placebo because the patients did not know who was being prayed for and who wasn't. Nor can they be attributed to social support, because there was no interaction of any kind between patients and those who prayed for them. If the results cannot be attributed to a placebo effect or to social support, both of which would have been worth investigating in their own right, what then? The results of this study violate two of the most fundamental assumptions of biomedical science: 1) any cure for physical illness must be found in matter (prayer doesn't fall into this category and couldn't possibly have an effect); and 2) God does not exist, and therefore praying to God couldn't possible work. The results of this study imply (but do not prove) violations to both of these doctrines – there might be a God involved and there might be an energy source involved in the healthier outcome of the prayed for group.

The results of this study were initially ignored. Then there was a follow-up study, also of coronary patients [1.5] that further supported the beneficial effects of intercessory prayer. That study, published in a more prestigious journal, was followed by a backlash of commentary and an article [1.6] declaring that, "no effect of intercessory prayer has been proven". A subsequent trial [1.7] that did not give the intercessors any feedback on the progress of the patient being prayed for, found no significant improvement in prayed for patients. An additional trial [1.8], randomly assigned the cardiac patients to three Christian sites that had agreed to pray for the patients for 14 days each. The agreement was that the daily prayer assignment would be covered by someone at the site. No feedback was given on patient progress and the fact that the lists could theoretically be prayed for by different people every day, lessened the probability that the prayer would feel connected to the prayed for person. This study also failed to show any improvement in patients who received prayer over those who did not [1.9]. A review of the literature on intercessory prayer in general stated that "although some of the results of individual studies suggest a positive effect of intercessory prayer, the majority do not and the evidence does not support a recommendation either in favour or against ..."[1.10].

The reason that feedback about the patient may be important for intercessory prayer has to do with the energy associated with prayer. When we pray for a loved one, we pray with love, emotion and intensity. When a stranger is being prayed for, progress reports help the intercessor feel connected to the person and to the results of the prayer. If progress is good, the intercessor may continue to pray in the same manner. If the patient's condition worsens, the prayer may be intensified, made more specific or changed in some other way. Why does this matter? Since the measurement of brain waves essentially involves measurement of energy fields on the surface of the scalp, it is not illogical to assume that thoughts have energy. In fact we can measure electrical potentials that are evoked by the brain's reaction to certain stimuli. So why would thoughts involved with prayer not also have energy? If these thoughts (prayers) involve a force field, then the possibility exists that they could influence other force fields (e.g., heart wave frequencies). The mechanism of influence (movement of the field through space/time, non-locality, etc.) is yet to be elucidated. However, these questions are not as strange as they seem and will be explored more fully in the section on the physics / biology interface. What the example illustrates is the way that medical science is influenced by its belief systems. In this case, the procedure of the successful intercessory prayer study was ignored by researchers who did not understand or believe that it mattered. When new trials using different methodology failed, the assumption was that intercessory prayer didn't work.

Any discussion of matter as it relates to the nature of reality would be remiss if it did not include a discussion of current theories concerning human evolution. Since Darwin's theory of evolution is currently the dominant theory in biology, and since it was derived during the period when materialism was consistent with theories of physics, this is a good place to examine its tenets.

1.2. DARWINIAN EVOLUTION

"...history will ultimately judge neo-Darwinism as a minor twentieth-century religious sect within the sprawling religious persuasion of Anglo-Saxon biology."

—Lynn Margulis, Distinguished Professor of Botany [1.11]

"We conclude – unexpectedly – that there is little evidence for the neo-Darwinian view: its theoretical foundations and the experimental evidence supporting it are weak."

—H.A. Orr and J.A. Coyne [1.12]

As the above quotes illustrate, Darwin's theory of evolution is not without its scientific critics. The term 'neo-Darwinism' is sometimes used to describe the evolutionary theory that evolved from Darwin, even though at the time, genes as such, had not yet been discovered. What this theory postulates, is that the evolution of life started from inanimate matter which somehow evolved into a living cell and from there to complex organisms through a process of natural selection based on random mutations. It began when a single cell or molecule randomly found itself in an enzyme solution, at the right temperature, and by some mechanism (not fully elucidated but maybe involving lightning) developed the ability to start reproducing. Thus, through some as yet unexplained process, inorganic matter was transformed to organic (living) matter. Over time, random mutations occurred, and the ones that survived, somehow diversified, mutation by mutation, into the enormous number of species that currently exist, gradually evolving into human beings with our current level of consciousness. There were no general principles involved and thus, no design, because life began and evolved as an accident. The issue under debate in this book and in other areas of biology is not whether evolution happened because the extant data indicate that it did. The issue is also not whether the "fitness" of certain random mutations may have facilitated some aspects of survival. The issues addressed here are whether "survival of the fittest random mutation" was the primary driving force in evolution and whether the origin of life can be traced to inanimate matter.

Because neo-Darwinism has had such a tremendous impact on so many areas of science, it is important to examine its tenets closely. There are several problems with these theories that have been elucidated in a book by Michael Behe [1.13]. The first, is that the number of *random* mutations it would have taken to produce all the forms of insect, fish, bird and animal life that have evolved, would have required more time than has elapsed since the beginning of life on this planet. Relying on random mutations, the diversity of eye types (eye stalks on caterpillars, multifaceted eyes on flies, human eyes), body types, colors, legs, wings, lungs, gills, stomachs, bones, brains, etc. and the incredible number of insects, mammals, reptiles, birds, and fish on the planet would have taken an incredibly long time, even if certain genes are responsible for similar variations across species (which data now indicate is the case).

This is especially true because as organisms developed, their ability to defend against random mutations also increased as a mechanism to maintain equilibrium and survival.

Biologist Lynn Margulis challenges her fellow biologists to identify even one species (or for that matter, one biological organism) that has evolved gradually through random mutations, during the entire time that scientists have been keeping record. According to Margulis [1.14-1.15], they can't because there aren't any. She points out that there are, however, numerous documented cases of species emerging through symbiosis. This is a process whereby two organisms that once were part of a symbiosis, merge and form a new, more complicated organism. For instance, amebas that have become infected by bacteria incorporate these bacteria to form new intracellular organelles. This theory called "endosymbiosis," although originating much earlier, was introduced into modern biological thinking by Margulis in 1981 [1.16] and is now well accepted among biologists [1.17]. A classic example of this process in cellular physiology is the mitochondria that are responsible for cell metabolism in our bodies. These mitochondria have different DNA than the nucleus of the cell. According to one molecular geneticist [1.18] proto-mitochondrion entered the primitive eukaryotic cell (cell that has a nucleus) between two and three billion years ago. This bacterium at first retained the DNA of a free living organism but as the symbiosis matured, much of its DNA was transferred to the chromosomes. In contrast to Darwin's theory, which implies that evolution was a linear, step-by-step occurrence, whereby complicated species evolved through a long process of single mutations, Margulis explanation implies a nonlinear process.

Her nonlinear approach is also more consistent with the fossil evidence from the Cambrian "explosion" than is neo-Darwinism. What the data show, is that before the Cambrian period 543 million years ago, fossil evidence reveals only microscopic, vegetative types of organisms. Then "suddenly" during what is known as the Cambrian explosion, fossil evidence of every basic form of animal that we know today emerged [1.19]. Although this theory must be somewhat modified after the discovery of fossil evidence of a crustacean from the early Cambrian period [1.20], the tremendous diversification during that period still contradicts the concept of linear evolution. Although there has been an attempt to explain the tremendous diversification of species that happened as an 'S' shaped mathematical growth curve with a lag and then a 'log' phase of growth during replication, that reasoning is based on limited assumptions related to only two variables: available food and space [1.21]. It does not include the genetic changes required for the emergence of such a tremendous number of complex functional systems that supposedly came from random variation and natural selection. Although the mechanism of symbiosis probably does not fully account for what occurred during the Cambrian period, it is more consistent with the nonlinearity evident in evolutionary data than is neo-Darwinism. Many biologists examining the data from the Cambrian period agree with Darwin about common descent, but argue against natural selection as the driving force.

A third problem with the neo-Darwinian theory, is that it does not take into account the fact that there are many examples of spontaneous formation of complex morphological structures (snowflakes, oil droplets in water) that have nothing to do with mutations. According to Stuart Kauffman, a complex systems theorist, there are many systems that exhibit spontaneous order. Rather than supporting the theory of natural selection, this may actually contradict it. "If selection, when operating on complex systems which spontaneously exhibit profound order, is unable to avoid that spontaneous order, that order will 'shine through.' . . . Rather than reflecting selection's successes, such order, remarkably, may reflect selection's failure [1.22]."

Thus, the issue for many scientists who do not accept the post-hoc attribution of causality implied by neo-Darwinism, is not whether random mutations have occurred and survived, but whether there is evidence to support the theory that survival of the fittest random mutation has been the primary driving force in evolution. Although this debate is still ongoing in the biological research community, I believe the neo- Darwinists are fighting a losing battle. According to Kauffman, "It is not that Darwin is wrong, but that he got hold of only part of the truth [1.23]."

One reason that the issue of Darwinism is important is that it has also been applied to explain the evolution of human consciousness. The field of science responsible for these theories, is called sociobiology, and is an offshoot of neo-Darwinism. Sociobiology interprets human psychological evolution and aspects of mind such as feelings, cognitions, perceptions, conscience, etc., as the result of evolution by natural selection. Their argument is based on neo-Darwinism and on the fact that certain emotions seem to be universal to all human cultures, which is interpreted as their having evolved because they were necessary for survival. According to this line of thinking, feelings of love and altruistic gestures such as sacrificing personal gain for friends, family, or country, are really selfishly motivated in the sense that altruistically helping others promotes survival of one's own species. Anger, the opposite of altruism, is defended with the same argument, albeit reversed. Anger is universal to all cultures, it causes violence, violence causes death, the strongest and smartest do not get killed therefore anger contributes to the survival of the fittest. Thus, sacrifice of self for the common good, as well as survival of the most lethal warriors can both be explained by survival of the fittest. I don't think anyone would deny that altruism contributes to the survival of the human species, although I'm not sure there is such universal agreement with respect to anger/violence. The Holocaust in Nazi Germany and the genocides in Cambodia under the Khmer Rouge and in Africa under Idi Amin, all of which targeted intellectuals and artists as well as ethnic minorities, would seem to contradict the theory that survival of the most successful killers leads to survival of the smartest and "fittest" humans. Again, this is not to say that certain forms of cognitive functioning or emotions don't lead to survival. At issue here is the contention that emotions evolved as part of natural selection and that their sole function is to contribute to it. To assume that because feelings of altruism and anger have existed and survived down through the ages and in all cultures (at least as nearly as we are able to ascertain) is not sufficient reason to prove that humanity survived because of them or that they contributed to natural selection. This line of reasoning is referred to as a post-hoc attribution of causality. Just because I was standing in my neighbor's yard at 4:15 p.m. when their lawn sprinklers went on, does not mean that I caused them to turn on. Is it not possible that humanity has survived *despite* its aggression or is in the process of destroying itself because of it?

What these examples illustrate is that the 'objectivity' of scientists is relative. We require less evidence for a theory that fits our belief system than a theory that does not. In the case of sociobiology, the temptation to associate correlation (e.g., feelings and survival) with causality has overwhelmed the evidence. If our theories relating to the potential power of the human mind and the meaning and definition of consciousness are based on sloppy thinking, then science and the continued evolution of consciousness may be seriously impeded.

One of the main points made by biochemist Michael Behe in his book, "Darwin's Black Box," is that not only is the evidence supporting Darwinism's theory of natural selection weak, but the data gleaned from molecular biochemistry strongly support the opposite conclusion, namely that of intelligent design. He begins by presenting a number of examples of what he calls "irreducible design" in organic life. Irreducible design refers to a system that does not function at all and has no conceivable purpose, if even one single component is removed. His argument is that if Darwinism is correct, the system had to evolve bit by bit. That is, complex systems evolved because individual mutations contributed to survival of individual parts that eventually developed into systems, and these, in turn, contributed even further to survival. Behe claims that there are multiple systems in the human body whose individual bits had no purpose and could therefore not have contributed to the survival of the species, nor would they have been able to survive until the next part evolved through natural selection. In fact, in some instances the individual parts would have been destructive and would have destroyed the organism before the system had a chance to evolve. He exemplifies this with the blood-clotting system. It contains many components, including: fibrinogen, plasminogen, thrombin, protein C, etc. If one of these is missing, the whole system crashes. A system that develops gradually through natural selection is supposedly perpetuated because each mutation that survives has some intrinsic function that contributes to the survival of the organism. However, the components of the blood-clotting system have no function except in relation to each other. There would have been no reason for them to survive as separate entities. Margulis' theory of symbiosis, however, is consistent with evolution of whole organelles (rather than piecemeal evolution) but is unlikely to be a complete explanation for something as systemically pervasive as a blood clotting system.

Another example Behe uses to illustrate his point about irreducible design, is the intracellular transport system that moves things from one place to another within a cell. There are trillions of cells in the human body and each is surrounded by a membrane that protects what goes on inside of it from dissipating forces outside the cell. This means that it can perform functions such as storing nutrients, without interference from outside forces that might otherwise metabolize them. Cells with a nucleus, which are the kind we have in our bodies, are highly complex. They contain many subunits, called organelles that are also surrounded by membranes. The different organelles perform specialized functions within the cell. The nucleus of the cell contains the DNA, or genetic information. Mitochondria metabolize nutrient molecules and turn them into energy that can be used by the cell. Lysosomes degrade molecules that are no longer useful. The endoplasmic reticulum synthesizes proteins and the Golgi apparatus modifies them. In order for these organelles to do their job, there needs to be a transport system between them to carry things from one place to another within the cell. This requires an amazing amount of organization. The system must know what has to be transported, where it needs to go, and be able to supply a vehicle to transport it. If the transport system fails, it leaves a surplus in one area and a deficit in another. The single components of the system are not likely to have survived as single mutations because enzymes that have a constructive function in one area of the cell can cause devastation in another. *The system works as a whole, or not at all.* The enormous complexity of the cellular system can be summarized by the following quote from the biochemist Albert Lehninger:

> A bacterial cell synthesizes simultaneously perhaps 3,000 or more different kinds of protein molecules in specific molar ratios to each other. Each of these protein molecules contains a minimum of 100 amino acid units in a chain... Yet at 37°C the bacterial cell requires only a few seconds to complete the synthesis of any single protein molecule. *Not only can the bacterial cell make individual protein molecules rapidly, but it can make 3,000 or more different kinds of proteins simultaneously, i*n the precise molar ratios required to constitute a living functioning cell. [1.24]

Behe, himself a professor of biochemistry, summarizes the sum total of cumulative efforts from molecular biochemists investigating the cell in the following way:

> The result of these cumulative efforts to investigate the cell – to investigate life at the molecular level – is a loud, clear, piercing cry of "design!" The result is so unambiguous and so significant that it must be ranked as one of the greatest achievements in the history of science... The observation of the intelligent design of life is as momentous as the observation that the earth goes around the sun or that disease is caused by bacteria or that radiation is emitted in quanta. [1.25]

There is no contradiction between evolution and Divine design. The involvement of active intelligence in creation incorporates rather than contradicts the contribution of evolutionary principles such as natural selection, symbiogenesis, genetic drift, etc. However, the hypothesis of Divine design is that evolution was not random and that the principles were put in place by a Higher Intelligence. Based partially on the nature of reality implied by quantum mechanics, this book hypothesizes an additional principle to the ones already mentioned, namely that living energy systems, heretofore not mentioned in the context of evolution, have played a defining role.

Like Darwin's explanation of the principles behind evolution, this latter theory is an unproven hypothesis, albeit one with a good amount of supporting data. Subsequent chapters will provide evidence for the importance of energy in the growth and function of biological systems and indicate why the theory of a role for living energy systems in evolution is worth investigating.

For now, we must unequivocally conclude from the existing data that the principles of evolution are not fully understood. The following sections on emergent properties and complex systems will help to demonstrate the complexity of the issues involved.

1.3. EMERGENT PROPERTIES

> "....a chemical compound and its molecular environment may form a complex system in its own right, which exhibits emergent properties (e.g., solubility and lipophilicity) that are non-existent (and meaningless) at the level of description of isolated molecules, and which are part of the property space of the compound..."
>
> —B. Testa and A.J. Bojarski [1.26]

As illustrated in the previous section, an examination of the nature of matter and the theory of natural selection are inextricably intertwined with the way in which single cells evolve into complex systems. What we discover when we begin to examine this process is that the characteristics which "emerge" as individual molecules aggregate and form systems, are often so different from the properties inherent in the individual molecules that they cannot be extrapolated from, or predicted by an understanding of the nature of these component parts. Attempting to predict the characteristics of complex organisms on the basis of characteristics of individual cells is analogous to attempting to predict the characteristics of Shakespeare's plays or a book on U.S. patent law, by knowing everything there is to know about the individual letters of the alphabet. No matter how closely you examine the individual letters, you will not be able to predict the characteristics of a single one of Shakespeare's sonnets.

The biological equivalent is what happens when atoms come together to form molecules. In the formation of molecules, new properties emerge at the molecular level, which are completely different from those of the individual atoms. At the same time, some of the properties of individual atoms disappear [1.26]. This disappearance arises when the electrons in the outer shell of the atom (valence electrons), form chemical bonds with other atoms to create the molecule. This sharing of electrons creates constraints on the properties of the individual atoms, while at the same time, allowing new properties to emerge in the resulting molecule. These structural changes resulting from the chemical bonds between atoms forming the molecule are what partially account for the new characteristics that emerge [1.26]. An example of this is hemoglobin, the part of the blood that transports oxygen [1.27]. Hemoglobin is composed of four hemoproteins. When these hemoproteins combine to form hemoglobin, some of their original properties and functions, such as hydration patterns (ways of combining with water), are lost. But what emerges, is a system for transporting and delivering oxygen.

Another illustration of emergent properties is the characteristic of temperature [1.28]. The temperature of a liquid, solid or gas is defined as the *average* velocity (rapidity of motion) of all atoms of the substance being measured, and exists and can be conceptualized only on an aggregate level. Knowing the velocity of a single atom gives no information about the temperature of the aggregate substance and does not allow it to be predicted. This is because increasing the velocity of one atom or even several atoms may not cause any change in the temperature of the substance, since other atoms may simultaneously decrease in velocity. This, in fact, is the case when temperature remains constant. The emergent characteristics of these collective particles (i.e., temperature resulting from movement) cannot be predicted or understood even by knowing everything there is to know about the original atoms. It would not be possible to extrapolate an understanding of water temperature by even the most thorough examination of individual hydrogen and oxygen atoms constituting a water molecule.

This discussion of emergent properties is a good place to explain a little more about the state of "multiple potentiality" and how it works. Nothing about biochemistry is static. On the contrary, biochemical molecules are constantly in a state of flux, due to continuous interactions with their environment. There are certain kinds of molecules that can assume different three-dimensional spatial arrangements without breaking their chemical bonds, when parts of the molecule simply rotate around a single bond. These molecules are called *conformers* and the changes in their geometrical shape are an example of how fluctuation influences form. The shape of a conformer can be influenced by its surroundings. When in water, it will hide its "water avoiding" side-chains in a core and expose its polar groups to the solvent. Conformers also exemplify the influence of form on fluctuation, because their ability to fluctuate in this manner is dependent on their chemical composition (e.g., types of chemical bonds they have). The number of possible geometric forms the molecule can assume is called the *conformational space* [1.27]. This space is limited but has multiple possibilities. "The fluctuation of form and function generates a number of molecular states, which are snapshots of the molecule at a given point in time" [1.27]. This dynamic ability to change as the situation requires is an example of how dynamic biochemical properties are. What this means, is that the multiple possibilities expressed by the variation of form, function and fluctuation, as the molecule interacts with its immediate environment, lead to the expression of different characteristic states of the molecule, called properties. The range of possible properties or characteristics is called the *property space*. What this shows is that compounds interact with their environment, both influencing and being influenced by it. Their state at any given time is not just determined by the properties of their smaller subcomponents.

These examples illustrate an interesting aspect of emergent properties that contradicts the reductionist neo-Darwinian approach to evolution, namely their potential for dominating, at an aggregate level, what happens at a molecular level. This means that causality goes in both directions – from the molecule to the aggregate and from the aggregate to the molecule. Not only can the behavior of an individual atom or molecule influence its neighbors (i.e., the 'group'), but it can also be influenced by the aggregate. The rising temperature (increase in *average* velocity) of a liquid or gas influences the behavior (velocity) of individual atoms because the movement of groups of atoms or molecules influences other groups and, in turn, individual members of the groups. The ability of the aggregate to influence the behavior and properties of a single molecule, demonstrates the inadequacy of reductionism for understanding the complex systems that dominate our bodies and our environment. We simply cannot comprehend complex systems by examining only their smallest constituent parts and assuming unidirectional causation. Since both the human body and the human mind are highly complex systems, relying on a reductionist approach to understand them can result in highly erroneous conclusions.

Neo-Darwinism postulates that the complexity of the human organism is a result of random changes that evolved from a singularity (single cell). However, Schrödinger's equation states that a single state can only be achieved by the collapse of multiple potentialities. This implies the *exact opposite* of Darwinism. The physicist, Hans-Peter Durr, echoes Schrödinger when he says that the only way to understand singularity is to observe it in the context of the "whole" [1.3]. He illustrates this concept by the analogy of an embryo in the womb [conversation with HP Durr, 1999]. The blastula produced from the fertilized ovum contains the entire program for the unborn child. This "whole" first divides into sections representing different parts of the body. Then it subdivides into areas for different organ systems, and the subdividing continues to smaller and smaller levels, until specialized cells are formed, each containing the genetic 'blueprint' for the whole organism. To understand the meaning and function of a blood cell, you have to understand the body from which it is taken. If you knew nothing about a human body, you would never be able to extrapolate the blood cell's function by observing by it on a laboratory slide. The smallest constituent parts of the human body are derived from continuous subdivisions of the "whole." The "whole" becomes parts, which through their emergent properties form a new and different "whole," in a highly creative process. To return for a moment to our earlier question of what relevance quantum mechanics has for our daily lives, these examples illustrate how very important it is for the way we think about causality and the way we conduct biomedical research.

However, relatedness at the quantum level goes much deeper than the sharing of electrons by neighboring atoms. There is another aspect, namely the principle of "non-locality," which has even more profound implications. Although counter intuitive, this phenomenon has been verified in a comprehensive series of experiments. The best way to explain how non-locality works is to describe a typical experiment. There are a number of ways to design these experiments, but the fundamental principle is illustrated by the following [1.3]: a laser beam is directed at a stream of atoms, causing an atom to radiate "twin" photons with opposite polarizations, which are then propelled through a particle accelerator in opposite directions from each other, sometimes for miles. In each direction, Polaroid filters are erected. Each time one of these photons encounters a filter, it behaves as if it knows what the other photon is doing. We know this because when two photons are created in this manner, their angular momenta must always add up to zero, which means that they have opposite polarizations. Whether or not a photon gets through the filter, which is polarized in a particular direction, depends on its polarization at the time it encounters the filter. Whatever polarization photon "A" has, photon "B" must have the opposite. By knowing the polarization of filters encountered by the photons as they travel in opposite directions, it is possible to calculate whether the polarizations of the photons are coordinated. So whether photon "B" gets through its filter at angle "x", is "dependent" upon what happens to photon "A" encountering its filter at angle "y," (and vice versa), implying that the polarization of one photon changes in coordination with the other. The filters are in no way connected to each other and are separated by long distances (often miles) on different trajectories. Experimental evidence has repeatedly confirmed that each photon acts as if it were communicating with its "twin." It has also been demonstrated that the photons could not be communicating with each other without exceeding the speed of light [1.29], and that they are not pre-programmed [1.2]. How, then, is this possible? The answer is that we cannot explain it, except to say that the two photons are subdivisions of a "whole," not two separate entities communicating back and forth. It is as if the photons are similar to ocean waves that separate and travel in opposite directions but remain connected because they are part of the same ocean.

What these examples demonstrate, is the poverty of reductionism as an explanatory principle for understanding the nature of reality. Although "mystical" is not a word that scientists are supposed to use if they want to be taken seriously by their colleagues, the deeper one delves into the laws of the universe, the more mystical the implications become. Regardless of vocabulary, these laws point consistently towards "oneness" of an almost unfathomable dimension. Whatever else these data imply, it is clear that the materialist reductionist methodology dominating most of biology and biomedical research is a grossly inadequate tool for understanding physiology and evolution. It is time to relinquish this scientific dinosaur in favor of a theory that creates consistency between physics and the life sciences. Is there a God? That question has yet to be answered, but data are stacking up on the side of a "master plan," and this question cannot be avoided forever. The important implications for science are that this issue is no longer one that should be limited to the purview of philosophers and theologians.

1.4. COMPLEX SYSTEMS

Emergent properties are characteristic of all complex systems and this is nowhere more evident than in the field of biology. The evolution from atoms and electrons to molecules and from molecules to macromolecules, cells, tissues, and organ systems, involves a complex process resulting in functions that cannot be reduced to the properties of their constituent parts. New mathematical tools have been developed to grapple with these issues and the field of nonlinear dynamics, with its subspecialty, chaos theory, has provided exciting insight into the functioning of these complex macro level systems. The term "dynamics" refers to the fact that there is continual change occurring, and "nonlinear" simply means that events do not occur in an additive manner. An example of a linear relationship would be a situation where every increase in number of cigarette packs smoked is associated with a proportional decrease in months of survival time in the smoker. A nonlinear relationship, on the other hand, can be exemplified by a drug that promotes health up to a certain dose, but when that dose is exceeded, becomes toxic and can result in death. Nonlinear dynamical methods have demonstrated that there are often complex temporal patterns in living systems that have important implications for function. Most of the systems in the human body, including neuroendocrine secretions, chemical processes, and the electromagnetic properties of the brain and heart rhythms are nonlinear. They are also dynamic in the sense that they are continually changing and adjusting to meet the needs of the body in each successive moment.

The subset of nonlinear dynamics called chaos theory provides interesting insight into important aspects of function in a diverse number of macro-level systems. Chaos theory describes the function of systems that lie somewhere between periodic (occurring in regular, repeating patterns) and random (never repeating) sequences [1.30]. In sequences of events where chaos theory applies, future occurrences are deterministic but long-term outcomes are unpredictable. This is not as contradictory as it sounds. Determinism stems from the fact that there are no "error" components in the equations and the rules for determining changes in the system are usually few and simple. The equation is explicit but the outcome is unpredictable because of its extreme dependence on initial conditions, for which there is never complete information. Due to nonlinearity, the tiniest disturbance to the system can create a major (nonlinear) change in the way it reacts, and could even cause the system as a whole to bifurcate (change suddenly) into two new trajectories. Try to picture a marble being nudged slowly along the top edge of a slanted roof. For every tiny nudge, it moves forward a tiny bit. But suddenly, the marble is given the same tiny nudge with only a very slight deviation in angle and it falls 18 feet to the ground. A tiny disturbance has given rise to a major reaction.

The problem of predictability can be illustrated by thinking about what happens to a drop of food coloring in taffy that is being stretched and pulled by a machine [1.31]. It is possible to predict where the food coloring will be shortly after it is dropped into the taffy, but as time progresses and the taffy continues to be pulled, the coloring disperses throughout the taffy and one loses track mathematically. Even though the particles of food coloring are known to be within a limited area, the whereabouts of each particle must be calculated separately and the precision of knowledge required for these calculations at each step of the way for each particle, is overwhelming. However, if we think of the piece of taffy as the space where each particle of food coloring is plotted against another particle positioned a fixed number of particles away, the average line of movement (trajectory) of these points will be more predictable than if they were moving randomly. According to chaos theory, many series of events that may at first appear random because they do not exhibit obvious repetitive patterns are found to have long range deterministic patterns. These patterns are constrained within a limited range of behavior, called the state space, sort of like the food coloring that is constrained to the taffy. The drops of coloring can move around within that limited space, but not outside of it. The importance of these long-range temporal patterns for biological systems has been demonstrated by the fact that their break down, either into periodicity or randomness, is associated with dysfunction [1.32].

The dysfunction associated with a breakdown of long-range patterns has proven to be a useful tool when applied to biological systems, where chaos theory has been utilized as a tool for determining a system's health. The non-linear dynamical fluctuations in many parts of the body are partially determined by geometrical structure. Many of the body's structures have what are called "fractal" dimensions. A fractal is an irregular structure that contains subunits within subunits, each of which resembles (but is not necessarily identical to) the larger scale structure [1.33-1.34]. Large branchings are repeated on decreasingly smaller and smaller scales. Concrete examples are the bronchi in the lungs, the arterial and venous systems, and the nerve fibers in the heart called the His-Purkinje system, that carry the nerve impulses from the pace maker to the parts of the heart that need to contract to pump blood. In fact, this fractional scaling can be found abundantly in nature. These fractal structures serve the function of being the fastest and most efficient methods of transportation between complex spatially distant structures. Fractal-like distribution networks are ubiquitous in nature. Examples include extensive surface areas of leaves, gills of fish, lungs, guts, kidneys, chloroplasts (plant cells containing chlorophyll), and mitochondria (internal components of our cells that generate energy) [1.35].

The reason these structures are important for understanding chaos theory is that their multiple scales indicate that their function is not dominated by a single frequency [1.36], as would be the case with single scale structures. Having only one scale would limit the behavior of the structure to a periodic, predictable pattern, and this limited response pattern would mean that the complexity of reactions required to respond to the multiple changes and demands occurring from moment to moment within the body, would not be possible. This is why the break-down of complex dynamics is an indication of dysfunction. It tells us something about a biological system's flexibility and ability to adapt. In order to be healthy, a system must be highly flexible [1.36] and this flexibility, or multiple potentiality for response, serves the important function of enabling the system to respond to the ever changing multiplicity of internal and external demands. The range of variation in biological signals that makes this possible often masquerades as randomness when viewed in short temporal segments, but emerges as patterns when viewed over a longer range. These long-range patterns are themselves often fractal, mirroring the fractal dimensions of the underlying structures. Such fractal patterns of fluctuation have been demonstrated at increasing levels of complexity from DNA [1.37] through cellular processes [1.38], to organ systems. The significance of this flexibility becomes clear when it disappears, e.g., when an organ such as the heart or the brain loses its flexibility of response and becomes "mode locked" into a specific repetitive pattern [1.39].

Heart disease, the most common cause of mortality in both men and women, can serve as an example. Although the heartbeat of a normal person seems to have a steady rhythm when viewed in short segments of the electrocardiogram (ECG), it varies in complex patterns over time. The electrophysiological signal that can be seen on the ECG is composed of both sympathetic (arousing) and parasympathetic ("calming") nervous input to the pacemaker of the heart that combines with other influences such as the structure of the heart, to determine the strength and frequency of heartbeats. Variability in this signal is necessary in order for the heart to be able to respond to the myriad demands continually being made by internal and external physical and psychological stimuli. When physical activity or emotional pressure increases, the heart must respond by increasing the supply of blood carrying oxygen and nutrients to the organs required to respond. During periods of relaxation, less action is required. In order for the heart to maintain flexibility and preparedness for the constantly changing demands, it must have the capacity for a broad range of responses. This is somewhat similar to the difference between marching in lock step with stiff legs and walking loosely, with slightly bent knees, over uneven terrain. The person who walks with slightly bent knees will be able to continually adjust the pressure of his/her step to the dips, obstacles and varying elevation encountered. The person marching in lock step is likely to stumble and fall and will certainly expend a great deal more energy to walk the same distance. For similar reasons, a reduction in the variability (and thus flexibility) of heart rate and a breakdown in nonlinear temporal patterns is an indication that the heart is not functioning properly [1.40-1.45]. One of the signs of a diseased heart is that it can only respond to incoming stimuli with a very narrowed range of frequency response, which does not give it the required flexibility for meeting the body's needs.

The implications of nonlinear dynamics have not yet been fully grasped by the life sciences community. What these dynamics imply is that multiple potentialities for future complex events exist on a macro level, just as they do on a quantum level, and that although previous events determine the probability of the occurrence of future events, there is a dynamic built into the system, such that sudden changes or new events can be incorporated and integrated into the response. As in quantum mechanics, where unpredictability is not a result of measurement error but an inherent property of the nature of reality, an element of unpredictability is inherent in complex systems. Indeed, multiple potentiality is what helps them to function. Homeostasis with respect to bodily systems is, therefore, an outmoded concept. A more accurate term for describing the dynamic system of "state spaces" that are constrained in their range of behavior, but allow flexibility of response, is 'dynamic equilibrium'. In a healthy system stasis does not exist. There is dynamic equilibrium which provides an ever-changing dynamical interaction with the environment (be it the fluid environment of a molecule or organ, or the external environment of a person).

I believe that this dynamic also applies to "healthy" theoretical systems, where stasis is represented by doctrine. The papal doctrine of Galileo's time was unhealthy because it created an atmosphere that put rigid constraints on scientific thinking. So much so, that the learned men of the day refused to look into Galileo's telescope to see the data that might change their view of planetary rotation because they didn't believe in the concept. Materialist reductionism has also assumed the role of a doctrine in biomedical research by excluding other approaches from the scientific purview. The human body is one of the most complex systems in existence, composed of many, many smaller complex systems. Trying to understand complex diseases, such as cancer or heart disease, by dissecting them down to their smallest constituent parts (e.g., genes, viruses, mutations, signaling pathways, receptors), is doomed to failure if this approach is not complemented by others that encompass an understanding of emergent properties and complex systems.

REFERENCES

1.1) Heisenberg W. Natural law and the structure of matter. As quoted in, Heisenberg W. *Across Frontiers*. Ox Bow Press (translation by Peter Heath), 1990, p. 115.

1.2) Walker EH. *The Physics of Consciousness* 2000. Perseus Books, New York, N.Y.

1.3) Durr HP. Complex reality: Differentiation of one versus complicated interaction of many as viewed by a quantum physicist. From "Information, interaction, emergence: A possible access to the conceptualorder of reality," a lecture given in Bielefeld, Germany, Sept. 1994. Max-Planck-Institute for Physics, Werner Heisenberg Institute, Munich.

1.4) Byrd RC. Positive therapeutic effects of intercessory prayer in a coronary care unit population. *Southwestern Med J* 1988;81:826-829.

1.5) Harris WS, Gowda M, Kolb JW, Strychacz CP, Vacek JL, Jones PG, Forker A, O'Keebe JH, McCallister BD. A randomized controlled trial of the effects of remote, intercessory prayer on outcomes in patients admitted to the coronary care unit. *Arch Internal Med* 1999;159:2273-8.

1.6) Hamm RM. No effect of intercessory prayer has been proven. *Archives of Intern Med* 1999;160:1872-1873.

1.7) Aviles JM, Wheelan E, Harnke DA, Williams BA, Kenny KE, O'Fallen M, Kopecky SL. Intercessory prayer and cardiovascular disease progression in a coronary care unit population: A randomized controlled trial. *Mayo Clinic Proc* 2001;76:1192-1198.

1.8) Dusk JA, Sherwood JB, Friedman R, Myers P, Bethea CF, Levitsky S, Hill PC, Jain MK, Kopecky SI, Mueller PS, Lam P, Benson H, Hibberd PL. Study of the therapeutic effects of intercessory prayer (STEP): Study design and research methods. *Am Heart J* 2002:143:577-584.

1.9) Benson H, Duske JA, Sherwood JB, Lam P, Bethea CF, Carpenter W, et al. Study of the therapeutic effects of intercessory prayer (STEP) incardiac bypass patients: a multicenter randomized trial of uncertainty and certainty of receiving intercessory prayer. *Am Hear J* 2006;151:934-42.

1.10) Roberts L, Ahmed I, Hall S, Davison A. Intercessory prayer for the alleviation of ill health. *Cochrane Database Syst Rev* 2007; 1:CD000368.

1.11) Margulis L: Kingdom Animalia: the zoological malaise from a microbial perspective. *Am Zool,* 1990;30:861-75.

1.12) Orr HA and Coyne JA. The genetics of adaptation: A reassessment. *Am Nat.* 1992;140:726.

1.13) Behe MJ. *Darwin's Black Box: The Biochemical Challenge to Evolution.* Touchstone, Simon and Schuster, New York, N.Y., 1996.

1.14) Margulis L. Words as battle cries—symbiogenesis and the new field of endocytobiology. *BioScience* 1990;40:673-677.

1.15) Margulis L and McMenamin M. Marriage of convenience: The motility of the modern cell may reflect an ancient symbiotic union. *Sciences* 1990;30:30-37.

1.16) Margulis L. *Symbiosis in Cell Evolution : Life and its Environment on the Early Earth.* San Francisco: WH Freeman and Company, 1981.

1.17) Okamoto N. and Inouye I. A secondary symbiosis in progress? *Science* 2005;310:287.

1.18) Wallace DC. A mitochondrial paradigm of metabolic and degenerative diseases, aging and cancer: A dawn for evolutionary medicine. *Ann Rev Genet* 2005;39:359-407.

1.19) Kerr RA. A trigger for the Cambrian explosion? *Science* 2002;298:1547.

1.20) Fortey R. The Cambrian explosion exploded: *Science* 2001;293:438-439.

1.21) Gould SJ. *Ever Since Darwin.* W.W. Norton & Company, Inc. New York, N.Y., pp 128-131.

1.22) Kauffman SA. *The Origins of Order: Self-Organization and Selection in Evolution.* Oxford University Press, New York, N.Y., 1993, pp. 29-30.

1.23) ibid. p.XIII.

1.24) Lehninger Al. *Biochemistry: The Molecular Basis of Cell Structure and Function.* Worth Publishers, Inc., New York, New York. 1975, p. 11.

1.25) Behe MJ. *Darwin's Black Box: The Biochemical Challenge to Evolution.* Touchstone, Simon and Schuster, New York, N.Y., 1996; 232-233.

1.26) Testa B and Bojarski AJ. Molecules as complex adaptive systems: Constrained molecular properties and their biochemical significance. *Eur J Pharm Sci* 2000;11(suppl.2):S3-S14.

1.27) Testa B, Kier LB, Carrupt P-A. A systems approach to molecular structure, intermolecular recognition, and emergene-dissolvence in medicinal research. *Medicinal Research Reviews,* Wiley, New York, N.Y., 1997; 303-326.

1.28) Popper KR and Eccles JC. *The Self and Its Brain: An Argument for Interactionism.* Springer Verlag, Berlin, 1977; 34-35.

1.29) Seife C. Spooky action' passes a relativistic test. *Science* 2000;287:1909-1920.

1.30) Denton TA, Diamond GA, Helfant RH, Khan S, Karagueuzian H. Fascinating rhythm: A primer on chaos theory and its application to cardiology. *Am Heart J* 1990;8:1419-1440.

1.31) Skinner JE. Low-dimensional chaos in biological systems. *Bio/Technology* 1994;12:596-600.

1.32) Peng C-K, Buldyrev SV, Hausdorff JM, Havlin S, Mietus JE, Simons M, Stanley HE, Goldberger AL. Fractal landscape sin physiology and medicine: Long-range correlations in DNA sequences and heart rate intervals. In: Nonnenmacher TF, et al (eds) *Fractals in biology and medicine.* Birkauser-Verlag, Basel. 1994; 55-65.

1.33) Goldberger AL. Is the normal heartbeat chaotic or homeostatic? *News Physiol Sci* (International Union of Physiological Science / American Physiological Society) 1991;6:87-91.

1.34) Goldberger AL. Nonlinear dynamics for clinicians: Chaos theory, fractals and complexity at the bedside. *Lancet* 1996;347:1312-1314.

1.35) West GB, Brown JH, Enquist BJ. The fourth dimension of life: Fractal geometery and allometric scaling of organisms. *Science* 1999;284:1677-1679.

1.36) Goldberger AL, Amaral AN, Hausdorff JM, Ivanov PCh, Peng CK, Stanley HE. Fractal dynamics in physiology: Alterations with disease and aging. *Proceedings of the National Academy of Sciences* 2002:99(suppl1):2466-2472.

1.37) Stanley HE, Buldyrev SV, Goldberger AL, Havlin S, Peng C-K, Simons M. Scaling features of noncoding DNA. *Physica A* 1999;273:1-18.

1.38) Havin S, Buldyrev SV Bunde A, Goldberger AL, Ivanov PCh, Peng C-K. Scaling in nature: from DNA through heartbeats to weather. *Physica A* 1999;273:46-69.

1.39) Peng C-K, Buldyrev SV, Hausdorff JM, Havlin S, Mietus JE, Simons M, Stanley HE, Goldberger AL. Non-equilibrium dynamics as an indispensable characteristic of a healthy biological system. *Integr Physiol Behav Sci* 1994;29:283-293.

1.40) Bhalla US, Lyengar R. Emergent properties of networks of biological signaling pathways. *Science* 1999;283:381-387.

1.41) Poon CS, Merrill CK. Decrease of cardiac chaos in congestive heart failure. *Nature* 1997;389:492-495.

1.42) Makikallio TH, Hoiber S, Kober L, Torp-Pedersen C, Peng CK, Goldverger AL, Huikuri HV. Fractal analysis of heart rate dynamics as a predictor of mortality in patients with depressed left ventricular function after acute myocardial infarction. *Am J Cardiol* 1999;83:836-839.

1.43) Makikallio TH, Koistinen J, Jordaens L, Tulppo MP, Wood N, Golosarsky B, Peng CK, Goldberger AL, Huikuri HV. Heart rate dynamics before spontaneous onset of ventricular fibrillation in patients with healed myocardial infarcts. *Am J Cardio* 1999l83L880-884.

1.44) Kalon KL, Ho MD, Moody GB, Peng C-K, Mietus JE, Larson MG, Levy D, Goldberger AL. Predicting survival in heart failure case and control subjects by use of fully automated methods for deriving nonlinear and conventional indices of heart rate dynamics. *Circulation* 1997;96:842-848.

1.45) Huikuri HV, Makikallio TH, Peng CK, Goldberger Al, Hintze U, Moller M. Fractal correlation properties of R-R interval dynamics and mortality in patients with depressed left ventricular function after an acute myocardial infarction. *Circulation* 2000;101:47-53.

♦

CHAPTER 2

IMPLICATIONS FOR BIOMEDICAL RESEARCH

2.1. THE DOCTRINE OF MATERIALISM

The implications of the foregoing facts concerning materialism and complexity for biomedical research are extensive. In the culture of modern medicine, it is accepted as self-evident that particulate matter is the only fundamental reality and that all "being" and all existing phenomena can be explained as manifestations of this primary substance. This assumption, which serves as the frame of reference for defining acceptable biomedical hypotheses, is not consistent with the principles of modern physics. The quantum theoretical experiments described in Chapter 1, demonstrate that at a subatomic level, the difference between waves and particles disappears and that experimental investigations of the exact same phenomena can produce greatly varying results depending on the design of the experiment. In the medical sciences, research on mechanism and function has focused almost exclusively on the particulate aspect, ignoring the wave characteristics. The result is that research in the area of theoretical and applied physics now proceeds with fundamentally different assumptions about the nature of reality than research in the fields of medicine and biology.

Although the principal reason for the current state of affairs is that most biomedical researchers are simply ignorant of the principles of quantum mechanics, there are also scientists who are familiar with these principles, but who nevertheless think that what happens at a quantum level is not particularly relevant to what happens on the macroscopic level of physiological functioning. They argue that quantum uncertainties "average out" on this level, and that Newtonian equations give pretty accurate predictions of physical events. This argument ignores the important implications of what quantum mechanics actually says about the nature of reality and why investigating only the particulate aspect will give an inaccurate reflection of the underlying essence of what is being examined. If the difference between matter and force filed disappears at a subatomic level, then there are no 'building blocks' of matter to which causality can be reduced. This means that the materialist frame of reference dominating medical science and driving its search for disease cures is fundamentally inaccurate. It indicates that we may be ignoring very important information that might contribute to therapeutic advances because we are inadvertently excluding important data and failing to ask the right questions.

2.2. GENES ARE INFLUENCED BY ENVIRONMENTAL FACTORS

The best place to see the consequences of this faulty approach is with genetics. Sequencing the human genome was a tremendous accomplishment and has provided scientists with an important tool for understanding the blueprint of human anatomy, disease vulnerability, and the evolution of species diversity. However, it has also lead to an over emphasis on searching for single gene causality and an oversimplification of the complex processes involved in chronic diseases. Heart attacks don't happen in healthy hearts and tumors can't grow when systemic immunity, tumor suppressor genes, DNA repair mechanisms and programmed cell death are functioning properly. The major chronic diseases actually evolve through the interaction of many genes and environmental factors over an extended period of time. Attempting to explain these complex processes by attribution to single genes or signaling pathways is overly simplistic and has not produced the hoped for return on investment with respect to chronic diseases such as cancer [2.1].

One of the reasons is that the DNA sequence in all bodily cells is essentially the same (with slight variation). What accounts for organ specificity, such as differences between the heart and the liver, is a series of little 'switches' that sits on top of the DNA. Together, these 'switches' are referred to as the epigenome. They have the ability to turn genes 'off' and 'on' and to increase or decrease their expression. The epigenome has been compared to the software that programs computer hardware (i.e., the genes) [2.2]. It has been learned that with the exception of a few diseases that are called 'monogenic' (caused by a single gene variation), most diseases manifest as a result of multiple genes interacting with each other and with multiple environmental factors. A person's genotype (i.e., the sequence of base pairs associated with specific genetic loci on the gene) tells us something about his/her vulnerability for specific diseases. However, with complex diseases like cancer and heart disease, whether a person with a vulnerable genotype actually contracts that illness is strongly influenced by interactions with the environment. Thus, gene expression is not always as static as eye color would have us believe. In many cases it varies throughout the lifespan, depending on multiple environmental influences. What mediates these influences is the 'software' or the epigenomic "switches" that enable many individual genes to have more than one function. The dynamic nature of gene expression is particularly important with respect to major chronic diseases because it is responsible for increasing or decreasing systemic susceptibility to disease manifestation.

This is very good news because it means that we have more control over our destiny than was previously believed. Whether a person actually becomes ill with a disease for which s/he has a genetic vulnerability depends to a great extent on the tissue microenvironment surrounding the genes (hormones, signaling pathways, transmitter substances, enzymes, etc.). The microenvironment, through its influences on the epigenome, has the power not only to "turn on" or "turn off" cancer causing oncogenes but also to turn off genes that fight cancer such as tumor suppressor genes. Most people do not know that the body has innate defense mechanisms for fighting cancer just as it has mechanisms for fighting disease pathogens like bacteria. When these systems become overburdened by multiple or chronic challenges, they become dysfunctional. The things we do and experience in our daily lives: the nutrients in our diet, behaviors such as smoking and exercise, pollutants in the air and in our food, and even our emotions (producing stress hormones or hormones associated with well-being) can influence the epigenetic switches and change gene expression.

Even though the search for "candidate genes" associated with specific diseases or conditions is important, a shift in emphasis is beginning to take place as a deeper understanding of the complexity of gene functioning emerges. Many researchers are turning their attention to the exploration of interactions between networks of genes and the ways that these networks are activated or inactivated by environmental factors. These complex interactions are not only important for understanding disease susceptibility but are especially important for understanding developmental trajectories from gestation through childhood to adulthood.

Epigenomic research is still in its infancy but already it is beginning to verify what we learned in Chapter 1, namely that most chronic diseases are complex and that there is multi-directionality to causal forces. Causality does not just begin with the smallest particles (e.g., DNA base pairs) and evolve into increasingly more complex tissues until we have an entire organism. Genes assert their influence by forming the blueprint for proteins, and the chemical configuration of protein molecules sets limits for which other molecules they can bind with to form more complex proteins, tissues and organisms, directing them to combine with some molecules and not with others. However, chemical changes in the complex proteins and tissues surrounding the genes precipitated by environmental factors, can initiate changes in gene function by influencing how the gene and other genes in the network are expressed. This can cause genes to stop producing certain proteins, start producing others, or to increase or decrease the amount of protein they produce. Thus, causal influences are at least bi-directional – from the part (e.g., gene) to the whole (e.g., human organism) and from the whole to the part. To make matters even more interesting, these changes in gene expression are also heritable, which means that they can be passed on from parents to their offspring.

Why is this important? It is important because our hypotheses concerning causality determine the treatment options we investigate. If we assume that a disease or condition is caused by the DNA sequence (e.g. structure) of a single gene, we may investigate ways to either 'knock out' that gene or in some way change its structure. If we are aware that transcription factors are involved, we may focus on specific transcription factors as the target for therapy. However, the more we begin to understand the complex interactions of these factors with other biological systems, the more likely we are to seek treatment methods that that are aimed at restoring balance within and between multiple, interacting systems. This means examining the context of the genes and signaling pathways to understand why they have become dysfunctional. Currently, it is common practice to use animals whose immune systems are genetically compromised (genetic 'knock-outs') in order to facilitate reliable tumor growth for cancer research. Genetic knock-outs also help us to understand the function of the gene that has been deleted. However, the fact that it is difficult to get tumors to grow in animals with fully intact immune systems should give us pause for thought. Since humans are rarely born lacking genes, this animal model does not mimic tumor initiation in humans, and gives biased information about how tumors progress because certain types of systemic resistance have been permanently blocked. Immune dysfunction caused by genetic deletion is irreversible. Immune dysfunction in humans, on the other hand, often results from epigenetic influences (precipitated by poor diet, lack of sleep, chronic stress, etc.) that can be reversed. When we limit the source of the data describing the system, we risk limiting the knowledge we will acquire in our attempt to find a cure.

Although there are a few alleles (gene variations), such as those coding for eye color or the genotype for Huntington's Disease that display simple Mendelian inheritance (i.e., a one to one relationship between a single gene and a trait or characteristic called a phenotype), it is the rule, not the exception that multiple genes interact with each other and with multiple environmental factors to determine the phenotype. Most importantly, gene/environment interactions involve dynamic processes that are ongoing throughout the lifespan. Diseases like cancer and cardiovascular disease take many years to develop – even in smokers. This implies that they involve a progressive process of dysfunction and that if caught in time, the process can be reversed. The really good news is that because much gene function does not involve unalterable characteristics that are fixed at birth, there is a lot we can do to prevent diseases for which we may have a genetic susceptibility. The life stage at which specific environmental factors play the strongest role varies widely and depends a lot on the confluence of other factors which are present at the same time. Some of them are active throughout the lifespan (e.g. nutrients in our diet) and others (certain medications or environmental pollutants) may have the strongest influence in early stages of fetal or brain development.

An example of the dynamic relationship between genotype and environment can be illustrated by research in monkeys that has revealed the influence of 'mothering' on gene expression. Research has shown that a particular form of the serotonin transporter gene is associated with an increase in alcohol consumption in female monkeys compared with monkeys that have a different form of the gene. However, this is only the case when the animals are reared in an environment that models parental absence or neglect [2.3]. How is this possible? The neurohormones associated with nurturance are completely different from the neurohormones associated with being motherless (a type of stress), and can influence gene expression in different ways. In fact, the regulatory pathways of nurturing vs. non-nurturing behavior in offspring and the way this behavior regulates changes in brain expression of genes influencing such traits as fearfulness and stress reactivity have been extensively studied in animals [2.4]. This has been done under very controlled environments to assure that other, non-related factors do not confuse interpretation. This research shows direct influences of maternal behavior on gene expression in the brain of their adult offspring [2.5].

Human research has demonstrated that a part of the genetic process which influences aging is very vulnerable to the effects of chronic stress. Cells in the body wear out and must be regularly replaced throughout life. There is a section at the end of chromosomes called a telomere which serves as a buffer protecting the chromosome while it is being replicated. The chromosome is often not capable of replicating all the way to the end, so the body's defense against genetic mistakes that could result from incomplete replication is to place multiple repetitive sequences at the end of the chromosome. This means that if the last few segments break off or the gene is not replicated all the way to the end, it will not affect transcription of the relevant base pairs because the chromosome is only losing nonessential repeats (a wonderful design!). The new chromosome will still have what it needs to properly program cell functioning. As the body ages, telomeres become shortened and these shortened telomeres are associated with age related diseases. It has been demonstrated that both the perception of being stressed and duration of stress are associated with shorter telomere length in the blood of healthy premenopausal women. Women with the highest levels of perceived stress had telomeres shorter on average by approximately one decade of aging compared to those in low stress women [2.6].

As these short glimpses illustrate, functional genomics is a fascinating area with implications we are only beginning to fathom. As a scientist, it is difficult not to feel an incredible sense of awe when confronted with the beauty and complexity of the interactions between the genome and its micro- and macro-environments that constitute the functional systems in our bodies. The thousands of molecular processes that are simultaneously ongoing in a single cell and the precision and accuracy of temperature, chemical ratios, and temporal sequence that must occur in a time span of only a few seconds [2.7], along with the number of cells in the body that are simultaneously performing these complex internal processes while at the same time interacting externally with each other, is almost beyond the ability of the finite human mind to grasp. Everywhere one looks, "design" seems to be blaring in neon lights. Survival of the fittest cannot explain many systems in the body (e.g., certain cellular transport mechanisms) because in an incomplete (still developing) form they would not have functioned at all or would have been destructive [2.8]. This is not proof of design. However, given the strong appearance of design, the data demonstrating that 'survival of the fittest' is not the only force behind evolution (see Chapter 1), and the lack of any data that contradict design, scientific objectivity prohibits summarily dismissing the concept simply because it doesn't fit the current (demonstrably inaccurate) paradigm of materialist reductionism that dominates biomedical research. Lacking data to the contrary, design must remain a viable hypothesis.

What will be emphasized in the remainder of this chapter is what research in genomics and epigenomics has demonstrated, namely that gene/ environment interactions are complex and cannot be partitioned into separate, independent components. This new knowledge has not yet been widely disseminated even within the scientific community, but is becoming more pervasive as the systems biology approach to medicine continues to grow [2.9].

Unfortunately, outdated concepts are often so ingrained in our thinking that they are difficult to relinquish. The use of mathematical equations in twin studies to model the 'heredity' of complex traits, conditions or diseases without actually measuring genes or environment is a classic example of an outdated concept that is still in use today. These methods are entirely built on the assumption that genes and environment are separate additive components that 'sum to 100', so that if one of them is high, the other is by definition low. These twin models calculate the similarity or dissimilarity of a phenotype by comparing twins that come from the same egg (identical twins) with twins that come from two separate eggs (fraternal twins). From this comparison they calculate 'heredity', based on assumptions surrounding the fact that twins from the same egg have identical genes while those from two eggs have only 50% of their genes in common. One of the underlying assumptions on which the equations are based (and one which we now know to be inaccurate), is that the genetic contribution is static and separate from the environmental components, and that adding the two together will explain variation in a phenotype between individuals. Knowledge has now advanced to the stage where we know that gene expression and thus, function is dynamic and often dependent on environmental input. Gene/ environment interactions are biologically so intertwined that trying to separate them into distinct components is artificial and non-workable. These dynamic processes can lead to multiple simultaneous influences on the same gene. Sometimes these influences are synergistic, sometimes antagonistic. The simple conclusion from all of this is that gene expression is non-linear and is not a static. It is not now, nor has it ever been 'additive' with the environment.

The good news is that if your mother died of breast cancer, it does not mean that you are destined for the same fate because of your genes. Many of the women who get breast cancer do not have an identifiable genetic risk and many who do have a genetic risk or vulnerability, do not get cancer [2.10]. You may well have inherited a genotype that increases your risk but there are things you can do to influence how those genes are expressed. We now know that genome wide changes in epigenetic control mechanisms are common to almost all cancers and that epigenetic changes also precede tumor manifestation [2.10]. This provides concrete support that environmental factors can influence cancer risk through their influence on gene expression. What is important to understand is that having a family history may increase risk but does not automatically seal your fate. Much research is currently being conducted on factors such as toxicants and diet that may interact with genes to increase or decrease cancer risk.

Because both cardiovascular diseases and malignant neoplasms (tumors) are known to evolve from complex interactions of multiple genes and environmental factors, the integration of multiple systems and levels of functioning play an important role in determining whether, and to what extent, genes are expressed. Since causality goes in multiple directions, attempts to reduce questions of mechanism to information from only one of these levels (in this case the genetic) will in most cases, not provide sufficient information for understanding disease etiology or enable us to find the most effective treatments.

2.3. THE GENOME AND COMPLEXITY

Let us stop for a moment to contemplate the genetic code. It is a string of single bases, combined with phosphates and sugars to form a helical structure (the bases on either side of the helix pair with each other according to certain rules). The chemical properties of these bases do not contain any of the characteristics of the complex systems they encode, any more than the letters of the alphabet contain the concepts of love or tragedy reflected in the sentences of Shakespeare's plays. The emergent properties of the components of different systems produced by base pairs in the DNA sequence are not present in the isolated constituent parts (e.g., the bases). They emerge from the interaction (and inter-relatedness) of the complex alleles (forms of a gene) composed of these sequences that interact with other alleles, with the epigenome, proteins, enzymes, hormones, transcription factors, tissues etc., all of which interact with the environment to form a blueprint for the function of complex organ systems. Just as the movement of many molecules forms the emergent property of temperature not present in single molecules and can, as a whole, influence the movement of a single molecule [2.11], so the tissues and molecules surrounding the genes can influence their function. Genes do not exist in a vacuum. Their ambient environment is a chemical soup that changes with the varying physiological state of the organism. The adaptability and redundancy of function created by these interactions is paramount to survival.

The current tendency of medical research to focus on genetics as the cause of complex diseases such as cancer or of personality characteristics such as depression is based on a reductionist belief that causality can be uncovered by breaking the system down to its individual parts and finding the one that is 'broken'. The problem is that the complexity of the phenotype cannot be found there. The genes function in a context and without the context we are like the blind man feeling the elephant's tail and describing the whole elephant as being 'thin like a snake'. If the scientific questions we ask are based on faulty premises and insufficient data, the answers we get are not going to be an accurate reflection of reality. All of the blind men feeling the different parts of the elephant are accurately describing what they are investigating, but their beliefs about the underlying reality of the elephant are based on incomplete data which causes them to misinterpret the results of their investigation. None of them has enough information to grasp an accurate concept of the essence of the whole elephant. Our current approach to complex disease is analogous. We focus on particulate aspects (e.g., genetics, genomics, proteomics) assuming that they will supply the key to understanding disease and behavior on a systemic level. However, in this instance, fitting together the parts of the elephant into a meaningful whole implies investigating not only the particulate but also the energetic aspects of the genome, as well as the emergent properties and systems that influence them.

2.4. GENES AND WAVE/PARTICLE DUALITY

In the chromosome, the DNA strand is wrapped around small "spool like" proteins called histones. This structure serves important functions, the first of which is to make the DNA as compact as possible so that it fits into the confined space of the cell nucleus. The second function is to keep the DNA from unraveling and being exposed to substances that could cause it to express prematurely or inappropriately. Since every cell contains the genetic program for every other cell in the body, we only want certain of those genes to be expressed at any one time in any one place. The structure of the gene allows for the important flexibility that is necessary for healthy organisms to survive – a dynamic ability to respond to the input and adapt to the demands of the surrounding environment, which includes factors such as hormones, nutrients, carcinogens, etc. These various substances can trigger processes which promote or prevent gene expression. When the body is functioning as it should, the gene is activated when and only when it is needed.

The beauty of this structure is that it stabilizes the DNA, protecting it from outside influences until expression is required. Contributing to this stability is the fact that the histone spool is positively charged and the DNA strand negatively charged, so that they adhere to each other like magnets. One epigenetic mechanism involved in chemical processes that allow outside factors to influence gene expression, is neutralization of the positive charge on the histone to dissolve the electromagnetic attraction with the negatively charged DNA, thus allowing it to unravel so that transcription factors can gain access to activate transcription.

However, the atoms in the DNA double helix are not separate and independent. They exist in clouds of identical electrons that do not belong to specific nuclei, but can only be said to have probabilities associated with multiple nuclei. The structure of the DNA helix is built on electrostatic attractions between the individual units, called bases, which are also dipoles (i.e., they are polarized like magnets due to asymmetry in their electron networks) [2.5]. These electrostatic forces are fundamental to function and have important consequences for sequence-specific interactions with proteins [2.12]. Similarly, proteins made by the genes contain coils in their structures called α-helixes, which function as "macroscopic" dipoles. These larger dipoles modulate the properties of several enzymes and define the process of coenzyme binding by numerous proteins [2.13-2.14]. Furthermore, protein folding and protein signaling are dependent upon dynamic energy states [2.15].

Thus, energetic forces are an inherent part of genomic and proteomic functioning. Unfortunately, the waveforms of inter-related potentialities (as defined by the Schrödinger equation) in these electrically charged particles are being totally ignored in current genetic and genomic research. If we reflect upon the fundamental meaning of wave / particle duality, namely that you determine whether you see a wave or a particle, by how you design the experiment, then half of the information contained in the genome is currently being ignored. We are setting up the experiment to see particles, not waves. In order to understand the importance of the wave form, you have to design an experiment that enables you to see it. It is quite probable that there is information in the waveforms associated with the DNA (a quantum coherent wave function – more on this in chapter 4) that cannot be grasped by pulling it apart and examining individual base pairs because the wave form is destroyed when the helix is dissected. This would be analogous to cutting out a piece of the heart and expecting the ECG patterns to be the same as those that were recorded in the intact heart. By ignoring the energy aspects of matter, we are neglecting a fundamental dimension of reality that is a prerequisite for a comprehensive understanding of the phenomena under investigation.

2.5. CANCER

2.5.1. Background

We have now discussed abundant evidence showing that reductionist approaches give us incomplete information and are not likely to provide sufficient knowledge for solving the complex interactions involved in the growth and metastasis of malignant neoplasms [2.10]. Despite tremendous effort and billions of dollars of investment, only marginal improvements in treatment outcome have been achieved since the 'war on cancer' was first declared. A recent review concludes that oncology has one of the poorest records for investigational drugs in clinical development [2.16] Cancer is still the second leading cause of death in the U.S. and is expected to become the leading cause of death within the next decade [2.17]. Estimated costs for cancer in 2010 total $263.8 billion, including $102.8 billion for direct medical costs, $20.9 billion for indirect costs of morbidity (e.g. lost productivity from illness), and 140.1 billion for lost productivity due to premature death [2.19].

A review of the Epidemiology and Population Studies (SEER) data base reveals that after being stable from 1975-1979, incidence rates in women for all cancers combined increased from 1979-2003 [2.19]. For men, there was an increase from 1979 through 1992, a decline from 1992-1995 and stable trends from 1995-2003 [2.19]. Mortality rates from cancer are higher among men than women and higher among blacks than whites. Survival rates obviously differ by type of cancer, some being higher than others. However, in general the mortality rates have decreased, albeit slowly. From 1990 – 2000, mortality decreased in amounts ranging from .08% a year to 1.8% a year [2.19].

What these statistics demonstrate is that although our knowledge concerning genetic mutations associated with malignancies has increased enormously, it has not translated into the level of therapeutic success that might have been expected given the huge economic investment. In 1984, the National Cancer Institute (NCI) launched a program that it claimed would halve the 1980 overall cancer mortality rate by the year 2000. Despite the fact that the NCI budget has increased 20 fold since 1971, the progress with respect to mortality rates has been modest. Given that success has been difficult to achieve, we might consider the possibility that important mechanisms are being ignored and that the animal models we are using may not adequately address tumor biology, etiology and formation.

2.5.2. *Tumor Progression and Immune Function*

Due to the fairly short cell cycle in almost every organ in the body, there is an ongoing process of cell replacement. Most cells wear out and must be regularly replaced, the number of days between replacement varying by cell type. The significant amount of cellular reproduction and repair that is necessary to keep the body functioning means that the probability for random mutations is extremely high. However, the body is equipped to deal with this. Natural killer cells "search and destroy" mutant cells, multiple mechanisms work to repair damaged DNA, tumor suppressor genes stop tumor growth and programmed cell death destroys dysfunctional cells. For the most part, this whole process occurs outside of our awareness. Only when something goes wrong with the multi-level system of safeguards, does a tumor emerge. In fact, the immune system is normally so strong, that it will reject an entire organ that is transplanted into the body, requiring people who receive transplants to take medicine to suppress the immune system for the rest of their lives. The belief that malignancy occurs every time there is a mutation, is a widespread misconception. It is important to emphasize that changes at a molecular level are often counterbalanced by systemic responses that correct molecular mistakes. In order for cancer to occur, there must be failure in multiple defense systems. This might be one reason why cancer, even in smokers, takes so long to develop. The body vigorously tries to repair the damage and is actually capable of reversing it if the smoker quits soon enough. It is possible for individual mutations, DNA adducts, or expression changes to be repaired, destroyed or reversed. However, there comes a point at which the burden becomes too great, overwhelming the systemic defenses. Repair mechanisms break down, signaling networks become dysfunctional and tumors develop.

Tumor development starts out benign, progresses to a stage where it is genetically unstable but with low invasiveness, and then ultimately to a metastatic phase, displaying increasing genetic changes [2.10]. As knowledge of this process has increased, it has become clear that there is no known single gene or genetic mutation that can account for tumor progression [2.20]. Genes cluster in nodes that are combined into nonlinear networks. A few genes have multiple connections with other genes but most have only one or two connections. Because of their structure and the fact that malfunctions usually occur randomly, dysfunction in a single gene has a low likelihood of disrupting an entire system. Gene networks aren't coupled in series like old fashioned Christmas tree lights, where if one failed, the system failed. This structural integrity may be one reason, along with built-in repair mechanisms, that tumor progression doesn't seem to start with a single mutation, even though mutations are involved in tumor progression. However, tumor formation and progression are characterized by multiple well known changes in epigenetic mechanisms (the 'programming software') that control how genes are expressed.

Together these facts are causing some scientists [2.1, 2.21] to think that we need to shift our perspective from seeing cancer as an entity to be destroyed, to regarding malignant neoplasms as the result of a process of dysregulation. The importance of this concept is that dysregulation can be reversed [2.21]. Thus, the treatment paradigm shifts from one that targets the killing of renegade cancer cells [2.22], to one that focuses on reversing the complex interactions involved in systemic dysregulation [2.23]. It is now known that tumor cells can be corrected by microenvironmental cues to revert to normal behavior [2.24-25]. Cell behavior (e.g., out of control proliferation) is seen as a result of regulatory imbalance and not as the initiating factor.

The concept of cancer as a process rather than a structural entity is also supported by research on cancers associated with well-known risk factors such as smoking. Smoking is a documented risk factor [2.26] that it is responsible for about 130,000 deaths annually from active and 22,200 from passive smoking [2.27]. However, although 80% - 90% of lung cancer patients are current or past tobacco smokers, only 10% - 15% of smokers actually develop lung cancer [2.28-29], and about 15% of men and 53% of all women with lung cancer worldwide (percentage of women is lower in the U.S), [2.30]) were never smokers [2.31]. These epidemiological findings strongly suggest that there are host differences in susceptibility to lung cancer and that the likelihood of a smoker developing lung cancer depends not only on how much and how long s/he has smoked but also on the presence of other etiological risk factors. Such factors include but are not limited to: 'at risk' genotypes [2.32-33], arsenic exposure [2.34-35], low socioeconomic status [2.36] and other environmental carcinogens.

There are no known single genes or genetic mutations that can account for the entire process of tumor initiation and progression, nor are there any known mutations that can account for metastasis or invasion [2.10]. There are however, multiple gene expression changes that have been demonstrated to contribute to tumor progression. Data suggest that more loss of tumor suppressor gene function may occur through epigenetically mediated gene transcription repression than via frank gene mutations [2.37-39]. This is exemplified by the epigenetic silencing of Hic-1 involved in modulating the activity of the p53 tumor suppressor gene [2.40-41]. Importantly, it has been demonstrated that epigenetic change precedes cancer and confers risk for cancer [2.20].

Two of the primary mechanisms that are involved in controlling epigenetic processes are DNA methylation (usually turns the genes off) and histone modifications (changes in the protein spools around which the DNA is wrapped that can influence its accessibility to transcription factors). Cancer is characterized by global changes in the epigenetic software that controls gene expression, including genome-wide reduction in methylation which can lead to the over expression of cancer genes (oncogenes). Data now suggest that even though tumors manifest at individual sites (e.g., lung, colon, breast), systemic changes have already occurred by the time they manifest [2.10]. These changes in gene expression can be responsible both for activating cancer genes and for down-regulating defense mechanisms such as tumor suppressor genes. These expression changes occur because of influences from the microenvironment in the cell which reflect and can be influenced by environmental factors originating outside the both cell and the body.

The immune system is also important in our defense against cancer. An interesting example of how robust our immune systems can be is illustrated by the Human Papilloma virus, which is known to be significantly associated with uterine cancer [2.42-2.44]. Because this virus is also associated with a common venereal disease, we know that it is fairly widespread and that not all women whose bodies are host to this organism, develop cancer. We also know that some women have the virus for many years and then "suddenly" develop cancer. Like the epigenetic mechanisms involved with lung cancer, the immune response in uterine cancer can apparently function successfully for years and then, for some reason, malfunction. The assumption of many lay people that the body is like a black box which automatically becomes ill when it is exposed to a virus or bacteria, is inaccurate. In fact we are often exposed to viruses and bacteria in public places and only seldom become sick. The reason for this is that our immune systems are amazingly flexible and robust, equipped to fight almost any type of bacteria or virus to which we are exposed. In most cases, we become ill only when the immune system is not functioning as it should, due to fatigue, undernourishment, chronic stress, biological strain from recent illness, or some other cause.

Understanding that complex systems are involved in tumor development has important implications for treatment. Our primary focus to date has relied heavily on the concept of pharmacological treatments and radiation that will kill the proliferating cells or inhibit specific signaling pathways. Unfortunately these treatments can also injure healthy ones, including those in the immune system that are needed to fight the cancer. Since cancer etiology is complex and clearly related to systemic malfunction, would it not make sense to also investigate the context, i.e., biological mechanisms related to the imbalance or dysregulation that allowed the abnormal cell proliferation in the first place? In other words, in seeking to find effective treatment for cancer, it seems that we should investigate not only the locally mutant signaling pathways or pathogens associated with tumor manifestation, but also the systemic malfunction in DNA repair, tumor suppression, immune and cell death mechanisms that failed in their normal function of defending the body against tumor initiation and progression. In order to do this, we also need to take into consideration the environmental context or demand overload that led to systemic malfunction. This approach has the potential for developing clinical interventions that prevent or reverse the processes that result in tumors, without destroying healthy cells.

2.5.3. *Cancer Treatment*

"Oncology has one of the poorest records for investigational drugs in clinical development with success rates that are more than three times lower than for cardiovascular disease [2.16]." Drugs used in the treatment of cancer can essentially be divided into those that target 'essential' functions and those that target non-essential functions [2.16]. An example of a drug that targets 'non-essential' functions is the estrogen-receptor modulator, Tamoxifen (an anti-hormonal), which targets specialized tissue (breast epithelium) that is not essential for life. Traditional cytotoxic pharmaceuticals target essential functions like cell division, which explains why their toxicity is not specific but affects multiple systems in the body. There are other drugs developed more recently that inhibit parts of key signaling pathways such as kinases (enzymes) [2.16] and are intended to be more specific and less destructive. An example is the drug, Imatinib which is used for chronic myeloid leukemia. Those that target upstream functions in signaling networks have less general toxicity than those that aim farther downstream. However, many of these drugs also affect other crucial survival and proliferation pathways whose destruction is deleterious [2.16]. The paradigm for all of these drugs is the assumption that cancer can be treated by targeting and either inhibiting or destroying a single function. This logic does not fit with what we know about gene-environment interactions and the paradigm has met with limited success.

Although modeled on the principle of antibiotics, which are designed to attack and kill bacterial invaders, this approach involves a fundamental fallacy when applied to malignant neoplasms. Unlike infection from an outside agent, the tumor is not a foreign invader but part of one's own body. What is toxic for the tumor is also toxic for other parts of the body. The term "side effects" applied to cancer therapies is not only a euphemism it is a misnomer. Toxicity is a main effect and the decision to categorize all tissue injury not involved with the tumor as "side effects" is purely semantic. Furthermore, it is an established fact that radiation treatment is associated with a significant increase in subsequent malignancy [2.45-2.48] and that some chemotherapies are also carcinogenic [2.49-2.52]. Although only about 5% of cancer deaths occur in people under the age of 45 [2.53], childhood cancers are one of the few areas where there has actually been a real improvement in survival. Unfortunately, survival after these treatments is often (but not always) associated with serious long-term side effects, such as endocrine and metabolic problems resulting from radiation and chemotherapy [2.54-2.55]. An overview of this literature indicates that the risk of later death due to causes other than recurrence of cancer is greatest in those children who were treated with a combination of chemotherapy and radiotherapy [2.54]. Indeed, in addition to being carcinogenic, chemotherapy and radiation are associated with a wide range of other toxicity, including effects on fertility, the cardiovascular system, nervous system, pulmonary system, blood and even death [2.56-2.67]. What is worse, data show that randomized clinical trials performed by the pharmaceutical industry often result in reporting that is biased in favor of the advantages of chemotherapy while downplaying the negative aspects, either due to selective reporting or to the publication of low quality papers that incompletely describe unfavorable results [2.68].

It is clear that despite a great deal of hard work, we still have a long way to go towards finding solutions to the treatment and prevention of cancers. As the information in this chapter indicates, cancer development is a complex process. Simplistic solutions targeting single mechanisms are unlikely to achieve more than a limited degree of therapeutic success. The following chapter discusses the implications that the energetic aspect of wave / particle duality has for biological and physiological functioning. This is an area that has not been pursued by traditional medical research because the traditional frame of reference emphasizes matter and sees no biological role for energy ('waves'). My hope is that the data in this book will demonstrate that it is time to broaden our frame of reference and to start asking different questions about the nature of reality and the etiology of complex diseases. The goal is to achieve success in finding and implementing effective treatment strategies for complex diseases such as cancer. For this we need a more accurate paradigm as well as integrated dynamical systems approaches to aid us in our endeavor.

REFERENCES

2.1) Knox SS. From 'omics' to complex disease: a systems biology approach to gene-environment interactions in cancer. *Cancer Cell Int* 2010, 10:11

2.2) Jirtle R. "Epigenetics" means what we eat, how we live and love, alters how our genes behave. [http://www.dukehealth.org/health_library/news/9322].

2.3) Barr CS, Newman TK, Becker ML, Lindell S, Shannon C, Champoux M, Lesch KP, Suomi SJ, Goldlman D, Higley JD. Interaction between serotonin transporter gene variation and rearing condition in alcoholpreference and consumption in female primates. *Arch Gen Psychiatry* 2004;61:1146-1152.

2.4) Meaney MJ, Szy M. Maternal care as a model for experience-dependent chromatin plasticity? *Trends in Neuro* 2005;28:456-463.

2.5) Weaver IC, Cervoni N, Champagne FA, D'Alessio AC, Sharma S, Seckl JR, Dymov S, Szyf M, Meaney MJ, et al. Epigenetic programming by maternal behavior. *Nat Neurosci* 2004;7:847-854.

2.6) Epel ES, Blackburn EH, Lin J, Dhabhar FS, Adler NE, Morrow JD, Cawthon RM. Accelerated telomere shortening in response to life stress. *PNAS* 2004;49:17312-17315.

2.7) Lehninger Al. Biochemistry: *The Molecular Basis of Cell Structure and Function*. Worth Publishers, Inc., New York, New York. 1975, p. 11.

2.8) Behe MJ. *Darwin's Black Box: The Biochemical Challenge to Evolution*. Touchstone, Simon and Schuster, New York, N.Y., 1996.

2.9) Hood L, Heath JR, Phelps ME, Biaoyang L. Systems biology and new technologies enable predictive and preventative medicine. *Science* 1004;3065:640-643.

2.10) Feinberg AP, Ohlsson R, Henikoff S. The epigenetic progenitor origin of human cancer. *Nat Rev Gen* 2006;7:21-33.

2.11) Popper KR and Eccles JC. The Self and Its Brain: An Argument for Interactionism. Springer Verlag, Berlin, 1977; 34-35.

2.12) Delcourt SG, Blake RD. Stacking energies in DNA. J Biol Chem 1991;23:15160-15169.

2.13) Sengupta D, Behera RN, Smith JC, Ullman GM. The alpha helix dipole: screened out? *Structure* 2005;13:849-855.

2.14) Hol WGJ. The role of the α-helix dipole in protein function and strture. *Prog Biophys Molec Biol* 1985;45:149-195.

2.15) Smock RG, Gierasch LM. Sending signals dynamically. Science 2009;324:198-2003.

2.16) Kamb A, Wee S, Lengauer C. Why is cancer drug discovery so difficult? *Nat Rev Drug Discovery* 2006, 6(2):115-120.

2.17) Stewart SL, King JB, Thompson TD, Friedman C, Wingo PA. Cancer morality surveillance – United States, 1900-2000. *MMWR CDC Sur veill Summ.* 2004; 53(SS03):1-108.

2.18) American Cancer Society. Cancer facts & figures 2010. Atlanta: American Cancer society.

2.19) Hayat MJ, Howlader N, Reichman ME, Edwards BK. Cancer statistics, trends and multiple primary cancer analyses from the Surveillance, Epidemiology, and End Results (SEER) program. *The Oncologist* 2007;12:20-37.

2.20) Feinberg A. Phenotypic plasticity and the epigenetics of human disease. *Nature* 2007; 447:433-440.

2.21) Schipper H, Goh C, Wang T. Shifting the cancer paradigm: must we kill to cure? *J Clin Onc* 1995, 13:801.

2.22) Dolgin E. Cancer metastasis scrutinized. Nature 2009, 461:854-855.

2.23) Hoeijmakers J. DNA Damage, Aging, and Cancer. *N Engl Med* 2009;361:1475

2.24) Kenny PA and Bissell MJ. Tumor reversion: correction of malignant behavior by microenvironmental cues. *Int J Cancer* 2003;107:688-695.

2.25) Leyi L, Connelly MC, Wetmore C, Curran T, Morgan JI. Mouse embryos cloned from brain tumors. Cancer Res 2003;63:2733-2736.

2.26) Burns D. Tobacco-related diseases. In: 2003: Elsevier; 2003: 244-249.

2.27) Gan Q, Smith K, Hammond S, Hu T. Disease burden of adult lung cancer and ischaemic heart disease from passive tobacco smoking in China. *Br Med J* 2007;16(6):417.

2.28) Alberg AJ, Samet JM. Epidemiology of Lung Cancer. *CHEST* 2003, 123:21S.

2.29) Mattson M, Pollack E, Cullen J. What are the odds that smoking will kill you? *J Am Public Health Assoc*; 1987; 77:425-431.

2.30) Subramanian J, Govindan R. Lung cancer in never smokers: a review. *J Clin Oncol* 2007; 25:561.

2.31) Parkin D, Bray F, Ferlay J, Pisani P. Global cancer statistics, 2002. *CA Cancer J Clin* 2005;55:74-108.

2.32) Sellers T, Elston R, Stewart C, Rothschild H, Vogler G. Familial risk of cancer among randomly selected cancer probands. *Genet Epidemiol* 1988;5:381-91

2.33) Tokuhata G, Lilienfeld A. Familial aggregation of lung cancer in humans. *J Natl Cancer Inst* 1963;30:289.

2.34) Guo H. Arsenic level in drinking water and mortality of lung cancer (Taiwan). *Cancer Causes and Control* 2004;15:171-177.

2.35) Hertz-Picciotto I, Smith A. Observations on the dose-response curve for arsenic exposure and lung cancer. *Scand J Work, Environ Health* 1993;19:217-226.

2.36) Ekberg-Aronsson M, Nilsson P, Nilsson J, Pehrsson K, Löfdahl C. Socio-economic status and lung cancer risk including histologic subtyping—a longitudinal study. *Lung Cancer* 2006;51:21-29.

2.37) Baylin SB, Chen WY. Aberrant Gene Silencing in Tumor Progression: Implications for Control of Cancer. *Cold Spring Harbor Symp Quant Biol* 2005;70:427-433.

2.38) Jones P, Baylin S. The fundamental role of epigenetic events in cancer. *Nat Rev Genet* 2002; 3:415-428.

2.39) Herman J, Baylin S. Gene silencing in cancer in association with promoter hypermethylation. *N Engl J Med* 2003; 349:2042-2054.

2.40) Chen WY, Wang DH, Yen RC, Luo J, Gu W, Baylin SB. Tumor Suppressor HIC1 Directly Regulates SIRT1 to Modulate p53-Dependent DNA-Damage Responses. *Cell* 2005;123:437-448.

2.41) Chen WY, Baylin SB. Inactivation of tumor suppressor genes: choice between genetic and epigenetic routes. *Cell Cycle* (Georgetown, Tex) 2005; 4:10-12.

2.42) Murthy NS, Mathew A. Risk factors for pre-cancerous lesions of the cervix. *Eur J Cancer Prev* 2000;9:5-14.

2.43) Viikki M, Pukkala E, Nieminen P, Hakama M. Gynaecological infections as risk determinants of subsequent cervical neoplasia. *Acta Oncologica* 2000;39:71-75.

2.44) Haverkos H, Rohrer M, Pickworth W. The cause of invasive cervical cancer could be multifactorial. *Biomed Pharmacotherapy* 2000;54:54-59.

2.45) Kollmannsberger C, Hartmann JT, Kanz L, Bokemeyer C. Therapy-related malignancies following treatment of germ cell cancer. *Int J Cancer* 1999;83:860-863.

2.46) Matsumoto Y, Sakai K. Radiation-induced cancers following radiotherapy. *Gan To Kagaku Ryoho* 1999;13:2015-2020.

2.47) Yokota T, Roppongi T, Kanno K, Tsutsumi H, Sakamoto I, Fujii T. Radiation-induced squamous cell carcinoma of the chest wall seven years after adjuvant radiotherapy following the surgery of breast cancer: a case report. *Kyobu Geka* 2000;53:1133-1136.

2.48) Hall EJ. Radiation, the two-edged sword: cancer risks at high and low doses. *Cancer J* 2000;6:10032-3795.

2.49) Marselos M, Tomatis L. Diethylstilboestrol: I, Pharmacology, toxicology and carcinogenicity in humans. *Euro J Cancer* 1992;28A:1182-1189.

2.50) Matsumoto S, Matsud H, Uejima S, Kurita T. Secondary leukemia following ultra high-dose chemotherapy with peripheral blood stem cell autotransplantation for refractory testicular cancer. *Nippon Hinyokika Gakkai Zasshi* 2000;91:687-691

2.51) Dollmannsberger C, Kkuzcyk M, Mayer F, Hartmann JT, Kanz L, Bokemeyer C. Late toxicity following curative treatment of testicular cancer. *Semin Surg Oncol* 1999;17:275-281.

2.52) Fields KK, Elfenbein GJ, Trudeau WL, Perkins JB, Janssen WE, Moscinski LC. Clinical significance of bone marrow metastases as detected using the polymerase chain reaction in patients with breast cancer undergoing high-dose chemotherapy and autologous bone marrow transplantation. *J Clin Oncol* 1996;14:1868-1876.

2.53) Hoel DG, Davis DL, Miller AB, Sondik EJ, Swerdlow AJ. Trends in cancer mortality in 15 industrialized countries, 1969-1986. *J Natl Cancer Inst* 1992;84:313-310.

2.54) Murray RD, Brennan BM, Rahim A, Shalet SM. Survivors of childhood cancer: long-term endocrine and metabolic problems dwarf the growth disturbance. *Acta Paediatrica Suppl* 1999;88:5-12.

2.55) Rose SR, Lustig RH, Pitukcheewanont P, Broome DC, Burghen GA, Li H, Hudson MM, Kun LE, Heideman RL. Diagnosis of hidden central hypothryoidism in survivors of childhood cancer. *J Clin Endocrinol Metab* 1999;84:4472-4479.

2.56) Sklar CA. Overview of the effects of cancer therapies: the nature, scale and breadth of the problem. *Acta Paediatric Suppl* 1999;88:1-4.

2.57) Spaide RF, Leys A, Herrmann-Delemazure B, Stalmans P, Tittl M, Yannuzzi LA, Burke KM, Fisher YL, Freund KB, Guyer DR, Slakter JS, Sorenson JA. Radiation-associated choroidal neovasculopathy. *Ophthalmology* 1999;106:2254-2260.

2.58) Maberley DA, Yannuzzi LA, Gitter K, Singerman L, Chew E, Freund KB, Noguiera F, Sallas D, Willson R, Tillocco K. Radiation exposure: a new risk factor for idiopathic perifoveal telangiectasis. *Ophthalmology* 1999;106:2248-2252.

2.59) DeSantis M, Albrect W, Holtl W, PontJ. Impact of cytotoxic treatment on long-term fertility in patients with germ-cell cancer. *Interl J Cancer* 1999;10:864-865.

2.60) Menkes DB, MacDonald JA. Interferons, serotonin and neurotoxicity. *Psych Med* 2000;30:259-268.

2.61) Toyofuku M, Okimoto T, Tadehara F, Sumii K, Imazu M, Yamakido M, Sueda T, Orihashi K, Matsuura Y, Hino T. Cardiac disease late after chest radiotherapy for Hodgkin's disease: a case report. *Jap Circ J* 1999;63:803-805.

2.62) Mulhern RK, Reddick WE, Palmer SL, Glass JO, Elkin TD. Neurocognitive deficits in medulloblastoma survivors and white matter loss. *Ann Neurol* 1999;46:834-841.

2.63) Briasoulis E, Froudarakis M, Milonis HJ, Poponis I, Constantopoulos S, Pavlidis N. Chemotherapy-induced noncardiogenic pulmonary edema related to gemcitabine plus docetaxel combination with granulocyte colony-stimulating factor support. *Respiration* 2000;67:680-683.

2.64) Iarussi D, Indolfi P, Galderisi M, Bossone E. Cardiac toxicity after anthracycline chemotherapy in childhood. *Herz* 2000;25:676-688.

2.65) Cairo MS. Dose reductions and delays: limitations of myelosuppressive chemotherapy. *Oncology* 2000;14(9 Suppl 8):21-31.

2.66) Marron A, Carratala J, Gonzalez-Baraca E, Fernandez-Sevilla A, Alcaide F, Gudiol F. Serious complications of bacteremia caused by viridans streptococci in neutropenic patients with cancer. *Clin Infect Dis* 2000;31:1126-1130.

2.67) Uhm HJ, Yung WK. Neurologic complications of cancer therapy. *Curr Treat Options Neurol* 1999;1:428-437.

2.68) Djulbegovic B, Lacevic M, Cantor A, Fields KK, Benntt CL, Adams JR, Kuderer NM, Lyman. The uncertainty principle and industry-sponsored research. *Lancet* 2000;356:635-638.

CHAPTER 3

THE PHYSICS / BIOLOGY INTERFACE

3.1. ELECTROMAGNETIC PHENOMENA IN BIOLOGICAL SYSTEMS

3.1.1. *Developmental Aspects*

The previously mentioned divergence of the biomedical conceptualization of reality from that of modern physics has profound implications also for the formulation of scientific questions and the variables that are selected for investigation in medical research. Because of their unspoken but axiomatic belief in materialism, biomedical scientists have systematically disregarded years of published data concerning electromagnetic phenomena in biological systems. They have relegated energy aspects of biological systems, such as piezo-electricity in living tissues, neural oscillators, the cardiac pacemaker, and brain frequencies, to the status of epiphenomena that have no intrinsic function. Although these systems provide accurate information for diagnosing the presence of disease, they are being ignored as possible mechanistic pathways. It should by now have become evident from the information relating to quantum mechanics, that focusing only on the particulate form of matter to the exclusion of the energetic aspects, is tantamount to relegating half of the relevant information to the status of "not applicable."

Electrical properties are inherent in all living systems and seem to play a fundamental role in their development and subsequent maintenance. This should not come as a surprise since the essence of chemical interactions is electromagnetic or electrostatic attraction, exemplified by the creation of a covalent bond when two atoms share an electron, or the creation of an ion when one atom takes an electron from another, creating an unbalanced charge in the electron depleted atom. Electrostatic forces of attraction between oppositely charged groups also stabilize many amino acids (components of proteins) in living organisms [3.1]. Accumulated research on the electrical properties of living systems provides convincing evidence that they are present from the very earliest stages of development. One early review [3.2] summarized a group of experiments demonstrating that developing systems have strong steady electrical currents running through them. These currents originate in and are emitted by the embryo, and seem to play a crucial role in fetal growth and development. It has been demonstrated that at a very primitive stage, when a chick embryo consists mainly of two flat "sheets" separated by a space, the upper sheet contains a long groove through which cells enter into the cavity on the way to forming almost all of the internal tissues. Strong steady currents pour out of this groove and return elsewhere through this upper sheet [3.2]. In an experiment where conductive implants were made in a group of chick embryos to reduce the endogenous current shown to be leaving the embryos, growth was stunted in 92%, compared to the control group that had non-conductive implants, in which only 11% demonstrated stunted growth [3.3]. Further measurement of voltage gradients appearing at specific stages of development inside chick embryos indicate that the internal electrical fields serve as directional cues for cell movements during development. [3.4-3.5]. Experiments on other species support this, showing self-generated electrical currents also in a type of amphibian, called Xenopus [3.6]. Current is present in the first cleavage [3.7] during embryo formation and further experiments have shown oscillations in intracellular free calcium that have the same frequency as this cleavage. An increase in calcium preceded the first cleavage, and mitosis (cell division) was at its height when intracellular free calcium reached its peak [3.8], indicating a possible mechanism related to electrical potential differences created by the flow of ions across a membrane. Also in these amphibians, it has been demonstrated that disrupting endogenous electrical currents resulted in disfigurement [3.9]. This has been further verified in another species, a type of salamander called axolotl [3.10], where ionic current also predicts hind limb development. Additional findings from work on spontaneous bioelectric activity during early development indicate that a type of transmitter of neural signals (GABAergic) comes fully to expression in hindbrain (cerebellar) and forebrain (neocortical) networks, if the level of ongoing excitatory activity is sufficiently high during development [3.11]. In sum, what these studies seem to indicate is that electrical fields play an important role in *directing* growth and specialization in the developing embryo.

The role of electrical fields in growth and development does not seem to be limited to animal species but has also been found in plants. Eggs from a type of seaweed called, Pelvetia, also have internal electrical currents [3.12]. This current includes a "pulse" component in addition to a "steady state" component, both of which enter the egg's growing tip and leave the rest of the embryo. The rate at which the egg grows in length is roughly proportional to the size of the steady current through it [3.12]. Together, these data imply that endogenous electromagnetic fields are present in all organic life possibly from inception.

Shi and Borgens [3.5] have stipulated five criteria to determine whether these fields have an intrinsic function or are simply epiphenomena. Intrinsic function would be indicated if the following criteria are met: 1) embryonic cells must be responsive to outside voltages within the range of magnitudes measured within embryos; 2) disturbance of these internal gradients of voltage by imposed outside voltages in the physiological range should result in developmental arrest or abnormality; 3) disturbance should be most profound at the embryonic stages when internal fields are present within the embryo; 4) since the internal voltages are spatially polarized during development, the form of change in the embryo produced by an artificially imposed field should be predictable based on its orientation relative to the embryo's orientation; and 5) any technique that reduces or eliminates an internal voltage gradient should lead to developmental arrest or retardation. According to existing data, all five of these criteria have been met. However, mainstream biologists and medical scientists are either unaware of or ignoring the implications.

3.1.2. Cellular Functioning

Given the above data, it is not surprising that electromagnetic fields continue to play a role in physiological functioning, even after birth. One important role is that played by electrical membrane potentials in cellular functioning. All animal cells are surrounded by a membrane and electrical potential differences between the inside and the outside of the cell serve many functions. It has been proposed by Rapp, et al, that cellular oscillations also play an important role in biochemical regulation by protecting the signal from distortion [3.14]. They use the analogy of analog to digital conversion, where encoding the signal to a frequency, protects it from the ambient "noise" of cellular processes while it is being transported through the cell. In this case, the analog signal can be in the form of a hormone attaching to a receptor or secretory cell, light attaching to a photoreceptor, or mechanical pressure attaching to a stretch receptor. The intensity of the incoming signal determines the frequency to which it will be encoded. After the signal travels through the cell to the response site, the frequency to which it was encoded determines the intensity of the analog response (e.g., hormone secretion, mechanical response, etc.) [3.14]. Membrane potentials can be measured in cells as diverse as those of the outer layer (cortex) of the adrenal gland [3.15] located near the kidney and of the pancreatic islet cells, where repetitive oscillations are induced in the beta-cell membrane when glucose and other metabolizable sugars elicit insulin release [3.16]. Another example of this "analogue / digital transformation" is the way that the cylindrically shaped cochlear outer hair cells function to transform acoustic energy into electrical energy in the cochlea [3.17]. In fact, an examination of the literature indicates that oscillations in biological systems are ubiquitous [3.18]. There are neural oscillators, cardiac oscillators, oscillations in secretory cells in the anterior pituitary, adrenal gland, pancreas and cardiac pacemaker.

3.1.3. Repair and Regeneration

One very interesting aspect of electromagnetic properties of certain biological systems is that they seem to play an important role in repair and regeneration. Experimental work in frogs and salamanders of different types has shown that there are natural electric currents emanating from severed limb stumps, which seem to play an important role in regeneration [3.18-3.22]. In experiments where adult animals were divided into three treatment groups: animals that were "sham" stimulated (received an implant that did not deliver electricity), those that received negative (cathodal) stimulation to the amputated limb, and those that received positive (anodal) stimulation, the results differed sharply. The group with the positive stimulation showed extensive degeneration. The sham group developed a disorganized skin (callus) layer, and the negatively (cathode) stimulated group had significant bone, muscle and nerve regeneration. Because the applied current never exceeded 0.2 μamp, the authors were convinced that the regeneration would not have occurred without the help of currents from within the body, which they were subsequently able to measure. Subsequent experiments also showed that blocking these naturally occurring currents that emanated from severed limbs prevented regeneration [3.21-3.22]. Similar findings have been made in "mystery" snails that have their eyes at the end of a stalk. Immediately following amputation, a persistent ionic current flows at the site and continues for 10 hours. This current then begins to decline and returns to baseline levels 60 hours after amputation, when the regeneration is complete [3.23-3.24]. Electrical fields have also been measured in the vicinity of epithelial wounds in isolated cow eyes, the polarity of which is more positive at the wound than away from the wound [3.25].

Higher vertebrates do not regenerate limbs, however, it has been demonstrated that imposing an electric field can improve neurological recovery. An example of this is the application of electric fields to promote axonal regeneration in spinal cords of adult guinea pigs after the cords were cut in half. (Axons are a part of the neuron that takes the nerve impulse away from the cell body toward another neuron.) In this experiment 25 percent of the experimental animals showed functional recovery but none of the control animals did [3.26]. In more refined experiments (also with guinea pigs), more than half of the 20 animals receiving applied voltage gradients showed a robust regeneration of severed axons and it was demonstrated that axon growth could be guided by an electric field [3.27].

Not unexpectedly, individual cells can also exhibit complex periodic behavior, both with respect to outgrowth and with respect to electrical activity [3.28]. This has been illustrated by other research on neurons. It has been demonstrated that under conditions of high neuronal activity, neurites (a type of axon) cease growing altogether, but also that different neurites have different thresholds for inhibitory activity. According to one view, neurons form networks only under the influence of internal electrical activity and growing neurons can best be modeled as neurite fields [3.29]. Earlier studies on neuronal growth using applied electric fields, demonstrated a robust phenomenon, called "overshoot," which involved a transient overproduction in connectivity [3.29]. What was learned from these studies about attempting to guide neurite growth with electric fields was that it involves working with a network and that the network is composed of cells with different thresholds. The spatial distribution of the cells allows for connectivity patterns that result in complex behavior. The neuronal network has emergent properties that do not necessarily reflect those of individual cells and these cells can also be influenced by the emergent properties of the network. Thus, the use of electric fields to stimulate functional neuronal growth is complicated. However, applied electric fields have been utilized successfully to improve functional recovery of severe acute spinal cord injury in paraplegic dogs, when compared with similarly injured, sham treated dogs [3.30].

Neurons and neurites are not the only types of cells that are dependent upon electrical potentials for growth and functioning. It has been known for some time that bone is electrically dynamic and that living bone drives an electric current through itself and into sites of damage [3.31-4.32]. It has been suggested that the ability of bone to heal itself after fracture may be mediated by electrical signals [3.33]. This is suggested by three phenomena, namely, the fact that short lived potential differences can be induced in bone by mechanical stress (the part under pressure is electronegative with respect to the part not being compressed); that steady differences in electrical potential are observed in living bone during growth, with the growing part being negative in comparison to the non-growing part; and by the fact that an injury to a bone also induces voltage gradients, which are again negative with respect to the uninjured parts. This knowledge has led to the use of varying electric fields in the treatment of fractures [3.31-3.5]. Although the mechanisms are still unclear, it has been demonstrated that pulsating electric fields influence bone cell proliferation [3.36-3.37] and gene expression [3.38-3.39] and that not surprisingly, the initial state of the target tissue is important in determining sensitivity to electromagnetic fields [3.40].

Together, these experiments indicate that electromagnetic fields coming from within the organism have important mechanistic roles in the growth, development and functioning of organic life forms. Furthermore, the use of applied electrical fields can facilitate neurological growth and healing. Some of this literature is decades old, and yet it has received scant attention from the biomedical research community. My hypothesis is that this lack of interest stems from the fact that these data are not consistent with a materialist concept of biological mechanisms. This literature has not been widely read because the topic is perceived to be irrelevant. Given the plethora of new research published daily and the limits on their time, scientists can only read a very small part of what is published. Naturally, they choose that which they consider to be most relevant. Anything dealing with electromagnetic fields is thought to involve epiphenomena and therefore not worth their time. Why else would something that appears to be so important for physiological functioning be so consistently ignored?

3.2. BIOLOGICAL RESPONSE TO AMBIENT ELECTROMAGNETIC FIELDS

If, as these data indicate, electromagnetic fields play an integral role in the functioning of organic life, is it possible that they can also be unintentionally influenced by ambient electromagnetic fields in the environment? This has become an important issue in recent years due to the increased use of appliances that generate electromagnetic fields and the power transmission lines needed to transmit electricity to homes. The average U.S. home with 60 hertz current has a magnetic field of 2 milligauss (mG) or lower, however the strength of this field can also reach 12mG from appliances such as electric razors and hair dryers, the sides of computer monitors, or appliances with large motors, such as refrigerators [3.41]. The question being asked is whether there is an effect from these fields on humans, and if so, whether it is positive or negative with respect to health. The other question is whether it is possible to manipulate electromagnetic fields and harness them in the treatment of human disease. As would be expected, the answers to these questions are not straightforward. Any effect depends on a number of factors, such as the strength of the field, the length of exposure, its combination with other exposures, and of course, the state of the organism being exposed.

3.2.1. Toxicity

Microwave radiation (2450 MHz, at 5-15 mW/cm²), which is non-ionizing and non-thermal, has shown a tendency to decrease white blood cell count in rats that were exposed for two hours a day, five days a week, in comparison to non-exposed rats [3.41]. A procedure which has been used to test for the detection of radiation exposure-induced genetic damage to cells, has also demonstrated that microwave radiation of high power for long exposure time (30 or 60 min), was able to cause genetic damage to cells of human white blood cells *in vitro* (i.e., cell samples in a laboratory) [3.42].

One of the most heated debates concerns whether or not residential proximity to high-voltage-power transmission lines is associated with increases in childhood cancer risk. Most of the data are epidemiologic in nature (i.e., observational studies in large populations) and although the researchers try to control for all relevant variables that could confound the data, this is difficult to do when you are not sure of the functional mechanism. A number of studies have reported significant increases in childhood leukemia with exposure to low level electromagnetic fields [3.43-3.47]. A pooled analysis of magnetic fields and childhood leukemia based on individual records from nine studies concluded that the 99.2% of children residing in homes with exposure levels < 0.4 microT (microtesla) had estimates compatible with no increased risk, while the 0.8% of children with exposures ≥ + 0.4 microT had a relative risk of leukemia that was approximately doubled, even after controlling for confounders [3.48]. One of the issues being debated concerns how the field exposure should be measured. One approach claims that the proper measurement of exposure involves "spot checks" of electromagnetic fields at the time of disease diagnosis, extrapolating these as indicators of the biologically relevant period of exposure [3.49]. The other approach advocates a calculation of the average field strength for the relevant time period of exposure. A careful study performed in Sweden, found that spot measurements could give very misleading results because the transmission line loads were so unstable [3.50]. Most lines measured in this study exhibited marked diurnal load-current rhythms during the year of measurement and six showed systematic weekday-weekend differences. In addition, during 1958-1985, a time period covered by some of these studies, average loadings of Swedish 220 and 400 kV power lines increased by about 1.3% a year. This means that contemporaneous load current measurements are not a good surrogate for historical averages. The importance of this issue is illustrated by one study using both types of measurements that found a significant association between magnetic field exposure using historical measurements but no increased risk with the "spot" measurement technique [3.51].

The issue surfaces again in studies of adult leukemia. Another Swedish study, designed to measure the effect of residential exposure to high-voltage power lines on leukemia in adults, did two types of measurements: calculations of the magnetic field at the time of diagnosis and calculation of cumulative exposure for the 15 preceding years. The first type of measurement (measurement at time of diagnosis) was not significantly associated with adult leukemia. The cumulative exposure index was marginally associated [3.52]. When occupational exposure to magnetic fields was added to the residential exposure, the risk for leukemia was tripled [3.53]. This study demonstrates the importance of including multiple sources of exposure in the calculation. An additional study of 408,000 people in Great Britain, living within a 10 km radius around a transmitter, found that the risk of adult leukemia within two kilometers of the transmitter was significantly increased and that risk decreased with distance from the transmitter [3.54]. A Canadian study of 31,453 Ontario electric utility workers also reported that the percentage of time spent above electric field thresholds of 20 and 39 V/m was predictive of leukemia risk even after adjusting for duration of employment, which was strongly related to risk. Those who had worked for at least 20 years and had the longest high-exposure levels had a risk approximately 10 times higher than normal [3.55]. However, an overview of the workplace exposure studies [3.56] comes to the conclusion that the data are still inconsistent although they provide some support for an association between magnetic field exposure and adult leukemia. The inconsistency may stem from lack of inclusion of other exposure sources in the calculations but this is still unclear.

This is further complicated by another methodological issue broached in these studies, namely whether selection bias with respect to subjects can be a confounder. One study [3.57] found that if it eliminated "partial participants," i.e., those that refused to allow access inside the home or those who only participated in part of the study, the association between acute lympho-blastic leukemia and high current configurations increased by 23%. Further research showed that 'partial' participants had lower socioeconomic status, which is often associated with worse health due to other factors. The issue of selection bias can become a particular problem in control groups, where mo-tivation to participate is lower.

Studies that have investigated the association between residential and occupational field exposure and breast cancer risk have been mostly negative [3.58-3.61], with the exception of one study [3.62] that found marginal significance for an association between these fields and breast cancer among estrogen receptor-positive women younger than 50 years at diagnosis. This group had a seven-fold increased risk but the range (confidence interval) was so large that it raises the question of whether there were other relevant varia-bles that might have narrowed the specificity to a smaller group had they been included. However a Russian report disagrees, concluding that the problem of electromagnetic field (EMF) exposure and breast, as well as certain other cancers, is real and warrants further research [3.63].

A snapshot of other types of cancer research and EMF exposure reveals a mix. There is one report of a positive association among electric utility work-ers in Ontario, between Non-Hodgkin's lymphoma and exposures above elec-tric field threshold intensities of 10 and 40 V/m [3.64]; no significant associa-tion between pancreatic cancer and EMFs [3.65]; and no increase in toxicity in cells growing in a laboratory, that were simultaneously exposed to a carcin-ogen and 60 Hz low frequency magnetic fields, over and above the carcinogen alone [3.66]. Although another study found a modest suppressive effect of magnetic fields on natural killer cell activity (the immune cells that attack cancer cells in the body) in one strain of mice, it did not find an increase in the incidence of neoplasia (tumor formation) [3.67]. However, the exposure in the latter study was relatively short and it is difficult to know whether it was sufficient to test an association with cancer.

There is one last area worth mentioning, though it is too soon to draw conclusions due to the small number of studies, and that is the area of the association between psychological disturbances and EMFs. One study on 138,905 male electric utility workers [3.68], showed an increased risk for sui-cide mortality related to years of employment as an electrician or line worker, and a dose-response gradient (i.e., the higher the dose, the higher the risk) between exposure to magnetic fields for the previous year and suicide mortali-ty. An additional study of 30,631 Danish utility company workers between 1978 and 1993, using estimated exposure levels to 50-Hz EMFs, reported an increased risk for senile dementia and motor neuron diseases compared to the general population but did not find an association for Alzheimer or Parkinson's disease [3.69]. However, so few studies have been done in this area, that any conclusions would be premature.

Judging from the above data, the evidence seems to be strongest for an association between residential and occupational exposure to EMFs and childhood and adult leukemia. There is skepticism in the mainstream medical community about the validity of these results because based on their view of plausible causality they cannot conceptualize a mechanism. However, given the data presented on the importance of endogenous bioelectric fields in fetal development and organ repair, and the effects of pulsating electric fields on biological function, I believe that this is an area that should be carefully monitored and researched.

3.2.2. Harnessing Electromagnetic Fields to Treat Disease

If, as we have previously stated, there exist endogenous electromagnetic fields in the body that are essential to its existence and function, then it is not illogical to wonder whether applied electromagnetic fields (EMFs) could be utilized to treat disease. Laboratory studies to date have supplied us with some hopeful indications. It has been demonstrated that a magnetic field can actually enhance the efficiency of DNA repair. Using a common strain of bacteria, otherwise known to cause stomach upset, it has been shown that DNA repair can be improved by short-term exposure to a 50 Hz magnetic field [3.70]. An experiment designed to investigate whether EMFs could influence immune system functioning, found that 0.1 mT, 60 Hz EMFs could induce a 20% mean-increase in the binding of T lymphocyte cells to T cell receptors (cells in the immune system) [3.71]. This study also showed that 60 Hz sinusoidal EMFs and a commercial bone healing EMF modulate signal pathways that regulate lymphocyte proliferation. The authors' conclusion was that their findings clearly illustrate that EMFs can regulate lymphocyte proliferation *in vitro* and *in vivo* (e.g., in the laboratory and in the body) and indicate that electromagnetic fields may provide an important tool for treating human inflammatory diseases. What should become apparent upon reviewing these data, is the possibility that exposure to the same frequencies that have been shown to be detrimental with chronic exposure, may have a positive effect when applied acutely in small doses.

In conclusion, there are now abundant data indicating that electromagnetic fields are important for embryonic development and human function. These data also suggest that environmental electromagnetic fields may predispose to disease or harnessed for healing. The clinical implications of the mechanisms described above are enormous and it is a wonder that they have not been a major focus of biomedical research. This bias needs to be corrected.

REFERENCES

3.1) Lehninger Al. *Biochemistry: The Molecular Basis of Cell Structure and Function.* Worth Publishers, Inc., New York, New York. 1975, p. 11.

3.2) Jaffe LF, Stern CD. Strong electrical currents leave the primitive streak of chick embryos. *Science* 1979;206:569-571.

3.3) Hotary KB, Robinson KR. Evidence of a role for endogenous electrical fields in chick embryo development. *Development* 1992;114:985-986.

3.4) Hotary KB, Robinson KR. Endogenous electrical currents and the resultant voltage gradients in the chick embryo. *Dev Biol* 1990;140:149-160.

3.5) Shi R, Borgens RB. Three-dimensional gradients of voltage during development of the nervous system as invisible coordinates for the establishment of embryonic pattern. *Dev Dyn* 1995;202:101-114.

3.6) Robinson KR, Stump RF. Self-generated electrical currents through Xenopus neurulae. *J Physiol* 1984;352:339-352.

3.7) Kline D, Robinson KR, Nuccitelli R. Ion currents and membrane domains in the cleaving Xenopus egg. *J Cell Biol* 1983;97:1753-1761.

3.8) Keating TJ, Cork RJ, Robinson KR. Intracellular free calcium oscillations in normal and cleavage-blocked embryos and artificially activated eggs of Xenopus laevis. *J Cell Sci* 1994;107(Pt 8):2229-2237.

3.9) Hotary KB, Robinson KR. Endogenous electrical currents and voltage gradients in Xenopus embryos and the consequences of their disruption. *Dev Biol* 1994;166:789-800.

3.10) Borgens RB, Rouleau MF, DeLanney LE. A steady efflux of ionic current predicts hind limb development in the axolotl. *J Exp Zool* 1983;491-503.

3.11) Corner MA. Reciprocity of structure-function relations in developing neural networks: the Odyssey of a self-organizing brain through research fads, fallacies and prospects. *Prog Brain Res* 1994;102:3-31.

3.12) Nuccitelli R, Jaffe LF. Spontaneous current pulses through developing fucoid eggs. *Proc Natl Acad Sci U S A* 1974;71:4855-4859.

3.13) Rapp PE, Mees AI, Sparrow CT. Frequency encoded biochemical regulation is more accurate than amplitude dependent control. *J Theor Biol* 1981:90:531-544.

3.14) Matthews EK, Saffran M. Effect of ACTH on the electrical properties of adrenocortical cells. *Nature* 1968; 219:1369-1370.

3.15) Matthews EK, O'Connor MD. Dynamic oscillations in the membrane potential of pancreatic islet cells. *J Exp Biol* 1979;81:75-91.

3.16) Ratnanather JT, Pope AS, Brownell WE. An analysis of the hydraulic conductivity of the extracisternal space of the cochlear outer hair cell. *J Math Biol* 2000;40:372-382.

3.17) Rapp PE. An atlas of cellular oscillators. *J Exp Biol* 1979;81:281-306.

3.18) Borgens RB, Vanable JW Jr., Jaffe LF. Bioelectricity and regeneration 1. Initiation of frog limb regeneration by minute currents. *J Exp Zool* 1977;200:403-16.

3.19) Borgens RB, Vanable JW, Jaffe LF. Bioelectricity and regeneration: large currents leave the stumps of regenerating newt limbs. *Proc Natl Acad Sci U S A* 1977;74:4528-4532.

3.20) Borgens RB, McGinnis ME, Vanable JW, Miles ES. Stump currents in regenerating salamanders and newts. *J Exp Zool* 1984;231:249-256.

3.21) Borgens RB, Vanable JW, Jaffe LF. Reduction of sodium dependent stump currents disturbs urodele limb regeneration. *J Exp Zool* 1979;209:377-386.

3.22) Jenkins LS, Duerstock BS, Borgens RB. Reduction of the current of injury leaving the amputation inhibits limb regeneration in the red spotted newt. *Dev Biol* 1996;178:251-262.

3.23) Bever MM, Borgens RB. Eye regeneration in the mystery snail. *J Experimental Zool* 1988;245:33-42.

3.24) Bever MM, Borgens RB. Electrical responses to amputation of the eye in the mystery snail. *J Exp Zool* 1988;245:43-52.

3.25) Chiang M, Robinson KR, Vanable JW. Electrical fields in the vicinity of epithelial wounds in the isolated bovine eye. *Exp Eye Res* 1992;54:999-1003.

3.26) Borgens RB, Blight AR, McGinnis ME. Behavioral recovery induced by applied electric fields after spinal cord hemisection in guinea pig. *Science* 1987;238:366-369.

3.27) Borgens RB. Electrically mediated regeneration and guidance of adult mammalian spinal axons into polymeric channels. *Neuroscience* 1999;91:251-264.

3.28) van Ooyen A, van Pelt J. Complex periodic behaviour in a neural network model with activity-dependent neurite outgrowth. *J Theor Biol* 1996;179:229-242.

3.29) van Ooyen A, van Pelt J, Corner MA. Implications of activity dependent neurite outgrowth for neuronal morphology and network development. *J Theor Biol* 1995;172:63-82.

3.30) Borgens RB, Toombs JP, Breur G, Widmer WR, Waters D, Harbath AM, March P, Adams LG. An imposed oscillating electrical field improves the recovery of function in neurologically complete paraplegic dogs. *J Neurotrauma* 1999;16:639-657.

3.31) Borgens RB. Endogenous ionic currents traverse intact and damaged bone. *Science* 1984;225:478-482.

3.32) Otter MW, McLeod KJ, Rubin CT. Effects of electromagnetic fields in experimental fracture repair. *Clin Orthop Relat Res* 1998;35(355 Suppl):S90-104.

3.33) Borgens RB. What is the role of naturally produced electric current invertebrate regeneration and healing? *Int Rev Cytol* 1982;76:245-298.

3.34) Satter SA, Islam MS, Rabbani KS, Talukder MS. Pulsed electromagnetic fields for the treatment of bone fractures. *Bangladesh Med Res Counc Bull* 1999;25:6-10.

3.35) Grace KL, Revell WJ, Brookes M. The effects of pulsed electromagnetism on fresh fracture healing: osteochondral repair in the rat femoral groove. *Orthopedics* 1998;21:297-302.

3.36) Fredricks DC, Neopla JV, Baker JT, Abott J, Simon B. Effects of pulsed electromagnetic fields on bone healing in a rabbit tibial osteotomy model. *J Orthop Trauma* 2999;14:93-100.

3.37) Hartig M, Joos U, Wiesmann HP. Capacitively coupled electric fields accelerate proliferation of osteoblast-like primary cells and increase bone extracellular matrix formation in vitro. *Eur Biophys J* 2000;29:499-506.

3.38) Heermeier K, Spanner M, Trager J, Gradinger R, Strauss PG, Kraus W, Schmidt J. Effects of extremely low frequency electromagnetic field (EMF) on collagen type I mRNA expression and extracellular matrix synthesis of human osteoblastic cells. *Bioelectromagnetics* 1998;19:222-231.

3.39) Loberg LI, Engdahl WR, Gauger JR, McCormick DL. Expression of cancer-related genes in human cells exposed to 60 Hz magnetic fields. *Radiat Res* 2000;153(5 Pt 2):679-84.

3.40) Muehasam DJ, Pilla AA. The sensitivity of cells and tissues to exogenous fields: effects of target system initial state. *Bioelectrochem Bioenerg* 1999;48:35-42.

3.41) Trosic I, Matausicpisl M, Radalj Z, Prlic I. Animal study on electromagnetic field biological potency. *Arhiv Za Higijenu Rada I Toksikologiju* 1999;50:5-11.

3.42) Zotti-Martelli L; Peccatori M; Scarpato R; Migliore L. Induction of micronuclei in human lymphocytes exposed in vitro to microwave radiation. *Mutat Res* 2000;472:51-58.

3.43) Feychting M, Ahlbom A. Magnetic fields and cancer in children residing near Swedish high-voltage power lines. *Am J Epidemiol* *1993*;138:467-468.

3.44) Feychting M, Ahlbom A. Childhood leukemia and residential exposure to weak extremely low frequency magnetic fields. *Environ Health Perspect* 1995;103 Suppl 2:59-62.

3.45) Feychting M, Schulgen G, Olsen JH, Ahlbom A. Magnetic fields and childhood cancer – a pooled analysis of two Scandinavian studies. *Eur J Cancer* 1995;31A:2035-2039.

3.46) Auvinen A, Linet MS, Hatch EE, Kleinerman RA, Robison LL, Kaune WT, Misakian M, Niwa S, Wacholder S, Tarone RE. Extremely low-frequency magnetic fields and childhood acute lymphoblastic leukemia: an exploratory analysis of alternative exposure metrics. *Am J Epidemiol* 2000;152:20-31.

3.47) Ahlbom A, Day N, Feychting M, Roman E, Skinner J, Dockerty J, Linet M, McBride M, Michaelis J, Olsen JH, Tyunes T, Verkasalo PK. A pooled analysis of magnetic fields and childhood leukaemia. *Brit J Cancer* 2000;83:692-698.

3.48) Bianchi N, Crosignani P, Rovelli A, Tittarelli A, Carnelli CA, Rossitto F, Vanelli U, Porro E, Berrino F. Overhead electricity power lines and childhood leukemia: a registry-based, case-control study. *Tumori* 2000;86:195-198.

3.49) Jaffa KC, Kim H, Aldrich TE. The relative merits of contemporary measurements and historical calculated fields in the Swedish childhood cancer study. *Epidemiology* 2000;11:353-356.

3.50) Kaune WT, Feychting M, Ahlbom A, Ulrich RM, Savitz DA. Temporal characteristics of transmission-line loadings in the Swedish childhood cancer study. *Bioelectromagnetics* 1998;19:354-365.

3.51) Feychting M, Kaune WT, Savitz DA, Ahlbom A. Estimating exposure in studies of residential magnetic fields and cancer: importance of short-term variability, time interval between diagnosis and measurement, and distance to power line. *Epidemiology* 1996;7:217-218.

3.52) Feychting M, Ahlbom A. Magnetic fields, leukemia and central nervous system tumors in Swedish adults residing near high-voltage power lines. *Epidemiology* 1994;5:501-509.

3.53) Feychting M, Forssen U, Floderus B. Occupational and residential magnetic field exposure and leukemia and central nervous system tumors. *Epidemiology* 1997;8:384-389.

3.54) Dolk H, Shaddick G, Walls P, Grundy C, Thakrar B, Kleinschmidt I, Elliott P. Cancer incidence near radio and television transmitters in Great Britain. *Am J Epidemol 1997;145:1-9.*

3.55) Villeneuve PJ, Agnew DA, Miller AB, Corey PN, Purdham JT. Leukemia in electric utility workers: the evaluation of alternative indices of exposure to 60 Hz electric and magnetic fields. *Am J Ind Med*;37:607-617.

3.56) Feychting M. Occupational exposure to electromagnetic fields and adult leukaemia: a review of the epidemiological evidence. *Radiat Environ Biophys* 1996;35:237-242.

3.57) Hatch EE, Kleinerman RA, Linet MS, Tarone RE, Kaune WT, Auvinen A, Baris D, Robison LL, Wacholder S. Do confounding or selection factors of residential wiring codes and magnetic fields distort findings of electromagnetic fields studies? *Epidemiology* 2000;11:189-198.

3.58) Forssen UM, Feychting M, Rutqvist LE, Floderus B, Ahlbom A. Occupational and residential magnetic field exposure and breast cancer in females. *Epidemiology* 2000;11:24-29.

3.59) Zheng T, Holford TR, Mayne ST, Ownes PH, Zhang B, Boyle P, Carter D, Ward B, Zhang Y, Zam SH. Exposure to electromagnetic fields from use of electric blankets and other in-home electrical appliances and breast cancer risk. *Am J Epidemiol* 2000;151:1103-1111.

3.60) Boorman GA, McCormick DL, Ward JM, Haseman JK, Sills RC. Magnetic fields and mammary cancer in rodents: a critical review and evaluation of published literature. *Radiat Res* 2000;153 (5 Pt 2):617-626.

3.61) Loberg LI, Engdahl WR, Gauger JR, McCormick DL. Cell viability and growth in a battery of human breast cancer cell lines exposed to 60 Hz magnetic fields. *Radiat Res* 2000;153(5 Pt 2):725-728.

3.62) Feychting M, Forssen U, Rutqvist LE, Ahlbom A. Magnetic fields and breast cancer in Swedish adults residing near high-voltage power lines. *Epidemiology* 1998;9:392-397l.

3.63) Grigor'ev IuG. Delayed biological effect of electromagnetic fields action. *Radiatsionnaia Biologiia, Radioecologiia* 2000;40:217-225.

3.64) Villeneuve PJ, Agnew DA, Miller AB, Corey PN. Non-Hodgkin's lymphoma among electric utility workers in Ontario; the evaluation of alternate indices of exposure to 60 Hz electric and magnetic fields. *Occup Environ Med* 2000;57:249-257.

3.65) Ojajarvi IA, Partanen TJ, Ahlbom A, Boffetta P, Hakulinen T, Jourenkova N, Kauppinen TP, Kogevinas M, Porta M, Vainio HU, Weiderpass E, Wesseling CH. Occupational exposures and pancreatic cancer; a meta-analysis. *Occup Environ Med* 2000;57:316-324.

3.66) Ansari RM, Hei TK. Effects of 60 Hz extremely low frequency magnetic fields (EMF) on radiation- and chemical-induced mutagenesis in mammalian cells. *Carcinogenesis* 2000;21:1221-1226.

3.67) House RV, McCormick DL. Modulation of natural killer cell function after exposure to 60 Hz magnetic fields: confirmation of the effect in mature B6C3F1 mice. *Radiat Res* 2000;153(5 Pt 2):722-724.

3.68) van Wijngaarden E, Savitz DA, Kleckner RC, Cai J, Loomis D. Exposure to electromagnetic fields and suicide among electric utility workers: a nested case-control study. *West J Med* 2000;173:94-100.

3.69) Johansen C. Exposure to electromagnetic fields and risk of central nervous system disease in utility workers. *Epidemiology* 2000;11:539-543.

3.70) Chow K, Tung Wl. Magnetic field exposure enhances DNA repair through the induction of DnaK/J synthesis. *FEBS Letters* 2000; 478: 133-136.

3.71) Nindl G, Balcavage WX, Vesper DN, Swez JA, Wetzel BJ, Chamberlain JK, Fox MT. Experiments showing that electromagnetic fields can be used to treat inflammatory diseases. *Biomed Sci Instrum* 2000;36:7-13.

CHAPTER 4

THE POTENTIAL ROLE OF BIOELECTRIC PHENOMENA IN THE TREATMENT OF DISEASE

We have now explored data showing that electromagnetic interactions are an integral part of many chemical bonds and are involved in the development and functioning of biological systems. Furthermore, biological systems have been shown to react both positively and negatively to applied and ambient electromagnetic fields. These data lead to a question not being addressed by traditional medicine, namely whether endogenous biological energy systems can be manipulated to aid in the treatment of human disease. The answer, I believe, is yes, and below I describe the personal journey that has led me to this hypothesis.

4.1. ENERGY TREATMENT - A PERSONAL JOURNEY

4.1.1. Acupuncture

Acupuncture is probably the oldest and most established form of the various energy treatment modalities. It began in China more than 2500 years ago and has been a part of traditional Chinese medicine ever since. It claims to manipulate human energy fields (called Chi meridians) in the treatment of disease. The theory behind it is that physical disease results from disharmony in these energy meridians which can arise from physical, emotional and spiritual causes.

The Chinese theory underlying acupuncture is that the harmonization and balancing of energy meridians is the mechanism through which healing of both localized and systemic disease manifestations occurs. This theory has, of course, been totally disregarded by the mainstream medical community in the U.S. because energy meridians make no sense in a frame of reference that deals only with particulate matter. According to the latter theory, there is no logical mechanism through which acupuncture could work, other than stimulating nerves that are connected to and can influence other systems. Therefore, in the west, acupuncture is believed to have clinical relevance primarily as an adjunct treatment in the relief of pain or the amelioration of nausea resulting from traditional cancer treatments. It would not occur to western scientists that acupuncture could be considered as a first line defense against cancer or problems with immune system functioning because they believe that there is no logical mechanism through which this could occur. Despite centuries of successful clinical use in China, funding for acupuncture research in the U.S. is only considered when research proposals can show that it functions according to mechanisms that adhere to western scientific beliefs, i.e., through something other than energy meridians (such as the nervous system or opioid systems). Since the inaccuracy of many aspects of this belief system has been clearly demonstrated, the theoretical underpinnings for this decision making process need to be challenged.

But let us continue our discussion by reflecting on some of the data. An overview of the literature indicates that acupuncture can be efficacious in the treatment of a broad range of conditions, including neurological problems, e.g., stroke rehabilitation [4.1], surgical anesthesia [4.2]; infectious diseases (e.g., chronic urinary tract infections [4.3], cystitis [4.4]); angina pectoris [4.5]; asthma [4.6], opioid dependence [4.7], immunomodulation in patients with malignant neoplasms [4.8-4.10], and the diagnosis of malignant tumors [4.11]. According to the 1997 Consensus development Conference on Acupuncture held at the National Institutes of Health [4.12], findings from basic research have begun to elucidate mechanisms of action, including the release of opioid and other peptides (which are building blocks of amino acids, from which proteins are synthesized) in the central nervous system and the periphery, as well as changes in neuroendocrine function. Examples of physiological factors that have been reported to be affected, include ACTH (a hormone secreted from the brain, especially during times of stress), oxytocin (also secreted from the brain when women breast feed, as well as during sensations of warmth and touch), vasopressin (secreted from the brain, constricts blood vessels), norepinephrine (a neurotransmitter), prolactin (also secreted from the brain, helps to facilitate breast feeding after birth), 17-hydroxycorticosteroids (secreted by the adrenal gland, helps to maintain blood pressure), as well as immune, and cardiovascular functions [4.13-4.15]. Because of these associations, the people reading this literature assume that these physiological changes reflect the causal mechanisms. But the assumption is being made without any serious thought to the causal pathways that would connect the needle locations to the hormone secretions and other effects. With the possible exception of pain, most of the points utilized by acupuncturists to treat a broad range of conditions do not connect with the physiological pathways that are hypothesized to be the curative mechanisms [4.16]. I believe that closer examination of the location of points in relation to the reported neurophysiological effects would indicate that the resulting changes are probably not causal but occur farther "downstream" in the causal pathway. Furthermore, acupuncture needles are not usually inserted into nerves. If they were, acupuncture would be extremely painful and it is usually fairly painless when performed by a skilled clinician.

When writing about scientific issues, I do not normally discuss anecdotal experience because it is not germane to the proof of scientific hypotheses. However, since the focus of this book is about examining how scientists' belief systems influence their hypotheses, and because my experience has influenced the scientific questions I am asking, I am going to make an exception and discuss several experiences that have contributed to my hypothesis that acupuncture should be investigated as a primary as well as a secondary treatment modality.

In 1994, I discovered two tumors that occupied an entire quadrant of my left breast. I should mention that I have avoided mammography screening because of the fear that even this minimal dose of radiation might trigger mutations in people who have a genetic predisposition for breast cancer, which I probably have because my mother died of it. I was exactly the same age as my mother had been when she discovered her breast cancer and the onset was uncannily similar to my mother's. Like hers, mine was premenopausal and large at the time of discovery. The tumor was in the upper outer quadrant of my breast, a site that develops more malignancies because of the increased tissue volume [4.17]. In addition to a family history of breast cancer, I had a number of other risk factors, including many years consumption of high strength birth control pills, late parity (birth of first child) and late menarche, as well as radiation exposure a few years earlier (almost daily chest x-rays during a compressed period of 10 days of hospitalization for pneumonia.

At the time I discovered these tumors, I was the single parent of a young child, whose father had died of cancer. She was still coping with that loss, and the thought of her also losing her mother to cancer was more than I could bear. I was determined that her childhood was not going to be scarred by the death of both parents. My first thought was to have a biopsy, but I knew that if the biopsy were positive, I would be under a great deal of pressure from the health care system to follow traditional treatment regimens, and I had already decided that whatever the diagnosis, I would not do chemotherapy or radiation (based on the data described in Chapter 2). I was convinced that my immune system and innate defense mechanisms had the potential to fight malignant growth and I wanted to strengthen that, not further compromise it as I believe chemotherapy would have done. If I did have cancer, the size and placement of the tumors suggested that it may already have metastasized and the existing data also indicated that under such circumstances chemo would not have had a high probability of success – perhaps a somewhat longer survival time but not a cure. However, the question of whether to have the tumors surgically removed was a different one. Benign or malignant, they were clearly not supposed to be there. According to research published shortly after I discovered the tumors [4.18-4.19], surgery to excise the tumor, an option I previously would have considered as a first step, was also risky. On the basis of laboratory studies, this research postulated that the rapid development of post-surgery metastases that often occurred and had previously been blamed on surgeons' failure to remove the entire tumor, was, in fact, being caused by loss of secretions from the primary tumor that had been removed. As contradictory as this seems to the lay reader, research has indicated that primary tumors secrete both promoters and inhibitors of angiogenesis (growth of vessels in a tumor that transport nutrients to help it grow). When the primary tumor is suddenly removed by surgery the secretion of angiostatin that previously had kept the growth of vessels in existing micro-metastases in check, stops and so the vessels in the existing metastases grow unhindered feeding their growth. I made a decision to temporarily postpone a biopsy because its purpose was primarily to determine the need for further treatment. Since I had decided not to pursue the traditional treatment methods, it served no useful purpose.

Given the risks, I decided that doing nothing might actually be less harmful than any of the previously mentioned options. However, I wanted to do everything in my power to heal, so I turned my attention toward other possible treatment modalities. My early reading of Werner Heisenberg's "Physics and Philosophy" [4.20] and subsequent readings on quantum theory, together with my reading of studies on electromagnetic fields in biological systems, had begun to coalesce into a hypothesis that illness first appears at an energy level and subsequently manifests on a physical level. I had formulated this hypothesis long before discovering the tumors in my breast. Although this theory was not proven (because it had not been tested), putting together data from different disciplines, I believed that there was strong evidence for its support. This hypothesis played a major role in my decision concerning treatment. Given all of the facts – the toxicity of traditional treatment modalities and data from biophysics – I opted for a method that would potentially treat the tumors on an energy level.

Although I had already made a decision to go with the hypothesis that God existed, I saw no reason to connect God with my current illness. In all likelihood, I had a genetic predisposition for cancer and I needed to find a treatment. I believed that an energy imbalance could reduce the effectiveness of the body's defense system and that a blockage in a meridian could be triggering tumor formation. From my interpretation of quantum physics and knowledge of biological electromagnetic fields, acupuncture simply made the most sense scientifically. Since I lived in Maryland, which has a rigorous licensing program for acupuncturists, I made a decision to try acupuncture as the first line of defense. At the time, other forms of "alternative medicine" such as herbal supplements, that function on a level of matter, in lieu of more traditional pharmaceutical products, seemed theoretically less compelling as primary treatment modalities. So, I went to an acupuncture clinic with a mixture of hope and skepticism. Patients were required to sign a form agreeing that they knew they were requesting an experimental, unproven treatment. I remember asking myself what the heck I was doing there, but a quick review of available options convinced me to forge ahead.

During my first interview at the clinic, the acupuncturist asked if I had had a biopsy. I told him no and explained the reasons. He told me that he would start with two treatments. If there was not a detectable change after these treatments, he wanted me to agree to a biopsy. This seemed reasonable and increased my confidence that he was not fanatically pushing his method of treatment to the exclusion of all others. After two treatments, there was a slight decrease in the size of the tumor. (At the time I still thought there was only one. Not until there was major shrinkage, did I feel the second one underneath.) The reduction influenced my decision to continue this course of treatment and to further postpone a biopsy. If it were malignant, I would not have changed my choice of treatment, and if it were benign, I would also have continued on the course of treatment until the tumors disappeared. I simply needed all my energy to heal and did not want to squander it fighting the health care system if a diagnosis of malignancy put me in the position of refusing their treatment options.

After the initial decrease, there was no change in tumor size during approximately two more months of weekly acupuncture treatments. However, after these two months, there was a sudden, very noticeable decrease in the size of the tumors. The subsequent decreases seemed to come in discrete steps. It would seem that nothing was changing for quite a while, and then there would be a very noticeable decrease in size. After approximately nine months of treatment, first weekly, then bi-weekly, the tumors were completely gone. I continued treatment about every six weeks on a preventive basis for several years and there were two recurrences. The first recurrence, in the same breast, required two months of acupuncture treatment on a more accelerated treatment schedule before it, too, disappeared. The second recurrence, in the other breast, required only two treatments before it disappeared.

At the time these tumors occurred, I had no thought of proving a scientific point or writing a book. If that had been my intent, I would most definitely have requested a biopsy and now greatly regret that I did not. Oncologists reading this book will be convinced that what I had was not cancer, but a benign fibroadenoma and they may be right. We will never know. However, whether they were malignant or not, it is interesting that acupuncture was associated with their disappearance. Of course, they might just have disappeared by themselves. Although I don't know of any data to support that theory, it is possible. Again, we will never know.

So let's continue the discussion with another anecdotal case of breast tumors, this one, diagnosed as benign in my daughter. At age 13, she developed a small lump in her breast that caused bleeding from the nipple. I discovered the blood while doing her laundry. When I asked her about it, she said that she had had the lump for about two months. Despite my previous tumor and treatment history, I took her to the family physician and was referred to a surgeon. The surgeon's diagnosis was "intraductal papilloma, probably benign." He recommended that the area be watched and re-examined every three months due to a family history of breast cancer (my mother's). If the mass did not enlarge, excision could be postponed to avoid deforming the breast before it reached maturity. The surgeon made it clear to me that the mass would not go away by itself and that it would have to be surgically removed.

At this point, I decided to take my daughter to the same acupuncturist who had treated me. He began talking to her and during their discussion, asked if she had received an injury to her big toe. She replied in the affirmative. It had been severely stomped by a horse as she was helping younger children with riding lessons at a summer camp and she had lost the nail. After an extensive discussion with her about her current activities and after examining the response to palpation in the front part of her legs, the acupuncturist smiled and told me that she would not require any acupuncture treatment if I would simply message two acupuncture points on the shins of her legs each night after she was in bed and before she went to sleep. He explained that the tumor was a blockage in an energy meridian that had been triggered by the accident to her big toe, and had been exacerbated by the dehydration she was experiencing in the frequent afternoon street hockey games that she played with her friends. Again, somewhat skeptical, but impressed with his earlier work, I did exactly what he said. After two weeks, the lump was entirely gone and has not returned in the intervening 14 years.

The next case involves a friend who developed a two-pound tumor that perforated her colon. She was working in a very rural area of South America at the time and suspected that the symptoms of stomach discomfort she was experiencing might be some sort of parasite. However, when it became severe, she returned to the United States for diagnosis and treatment. After routine x-rays revealed the tumor, she was immediately scheduled for surgery. The tumor was malignant and after removal, her oncologist insisted that she begin chemotherapy. After doing her own research, this woman too had come to the conclusion that chemotherapy would not increase her survival time and might actually shorten it. She refused both chemotherapy and radiation treatments, determined to find an alternative. Around this time, she and her husband moved to the Washington, D.C., area for his job. We met when she was referred to me by someone who knew of my tumor history and treatment.

When I called the acupuncturist who had treated me and told him about her case, he said that colon cancer was extremely serious and he did not know if acupuncture could help her. He said that he might even advise chemo. When she went for a consultation, he said that, based on her pulses (the method used in acupuncture and several forms of energy treatment to diagnose imbalances in the energy meridians), he was a lot more optimistic than her case history would have indicated. He began treating her with acupuncture, and included admonitions to consume absolutely no alcohol or cigarettes (she had previously been a light smoker). In the spring of 2000, she passed her five year survival mark and has subsequently remained cancer free. Of course, she might have just been lucky. Maybe she would have survived with no treatment at all. We will never know.

4.1.2. Jin Shin Jyutsu

The fourth anecdotal case is a woman who had thyroid cancer and was treated with Jin Shin Jyutsu. Jin Shin Jyutsu is a Japanese form of energy treatment based on a theory similar to the Chinese theory of acupuncture. Like acupuncture, it postulates energy meridians in the body that are associated with physical, psychological/emotional, and spiritual well-being and like acupuncture the method of diagnosis involves "listening to the pulses" (a method of feeling different pulses in the wrists). However, Jin Shin practitioners do not use needles. They treat the energy meridians by touching specific points on the meridian pathways with their hands (two points at a time to create a closed circuit) using the energy fields in their own bodies to help balance the circulation of energy. Two points are touched simultaneously and the theory is somewhat analogous to a "jumper cable". The practitioner forms a closed circuit between specific points on the patient and his or her own body. Jin shin practitioners hold their hands in the various positions until they feel a steady pulse between the points, then they move on to the next set of points. Different sequences are used for different types of problems. This method also includes self-help techniques that patients can do without a practitioner.

The patient, Joyce, was diagnosed with thyroid cancer (papillary cell) in September of 1992, she subsequently had both thyroid lobes surgically removed and during surgery the parathyroids were also accidentally removed. These glands control the amount of calcium in the blood and in the bones. As a result of that accidental removal, while still hospitalized, she experienced a dangerously low drop in calcium accompanied by cramping and an endocrinologist was called in to stabilize her electrolytes. There was very little success and as her health deteriorated, she continued to seek treatment at leading hospitals, ultimately going to Walter Reed Army Medical Center in 1995, because it was known for its expertise in this area. Monthly trips to Washington D.C. and a routine chest x-ray in 1996 showed a suspicious spot. Blood work determined that Non-Hodgkin's Lymphoma was present. This is a "waxing and waning" disease and the decision was made to "wait and see". The prognosis was estimated to be about 7 years of survival.

Over the ensuing years, she had five bone marrow aspirations and related tests for levels of white blood cells and platelets which showed that they were extremely low. During this period over 27 enlarged lymph nodes were identified, but because of the locations removal and biopsies were considered dangerous. Meanwhile, Joyce had intermittent contact with an old friend, Carol, who had become a Jin Shin Jyutsu practitioner.

In 1997, Carole went to visit Joyce and treated her on a daily basis for two weeks. During this time, Joyce who was quite ill experienced a marked improvement. Shortly after this visit, Joyce's husband and primary caretaker died suddenly and unexpectedly from a heart attack. Left in a house too large to care for in her weakened condition, Joyce sold the house and made a decision to move to Maryland and continue her treatment with Carole. Since the death of her own husband, Carole had been renting out one room in her house, so Joyce became the renter, which enabled her to receive Jin Shin treatment on a daily basis. She informed her doctors that she was receiving the treatment. Before long, Joyce showed marked improvement. Her metabolism and the medication she was taking to replace the function of her thyroid gland stabilized and most of the spots on the x-rays disappeared.

After a couple of years of continued Jin Shin treatment complemented with other alternative methods such as diet and supplements, during which Joyce had become healthy enough to lead a normal life, travel, and work as a stained glass artist, a large lump appeared on her hip. It was biopsied and found to be malignant. The diagnosis was B-Cell lymphoma. The tumor was surgically removed and Carole continued the Jin Shin. The wound healed rapidly and Joyce resumed her active lifestyle. More time elapsed and Joyce was diagnosed with another recurrence. She deteriorated rapidly and decided to have chemotherapy, a routine dosage of what she was told was a "new miracle drug" (Rituxan) for a series of four injections during four weeks. During this time the Jin Shin continued and Joyce once again became healthy. However, she subsequently developed cirrhosis of the liver. Joyce did not drink alcohol and the cirrhosis may have been the result of the chemo but despite hospitalization her condition deteriorated. She was diagnosed with end stage liver disease and given 3 to six years to live. That was six years ago. Carole intensified the Jin Shin and Joyce recovered within a short time. There have been several more crises of this nature but 15 years after she was told that she had approximately 7 years to live, Joyce is alive and leading a fulfilling life.

4.2. HYPOTHESIZED MECHANISMS

The reason I diverged from the discussion of the role of energy fields in biology, to relate several personal anecdotes, is that these experiences have influenced my thinking concerning physiological mechanisms. As I stated previously, the viewpoint of traditional medicine, is that if acupuncture works, it must work through the same pathways that allopathic medicine works. My breast tumors were primarily treated with needles in the feet and wrists. What neurophysiological or neuroendocrine pathways could those needles possibly tap into that would be related to breast tumors? My daughter's tumor was treated by messaging two points on her shins. If these were the only two cases of successful treatment in locations that didn't make sense from a traditional perspective, the answer would be obvious: acupuncture wasn't responsible for these cures, so they are irrelevant as examples. However, acupuncture efficacy has been documented in hundreds of studies where the needles do not tap into logical physiological pathways. An example is the treatment of opioid dependence [4.7] with auricular (ear) points. If one were trying to elicit a response in opioid receptors this would certainly not be the preferred target for intervention.

If neurophysiological and neuroendocrine changes occur but are not the cause, then what is the cause? Let us take a novel approach and examine the Chinese hypothesis – that the mechanism of healing involves energy meridians. This hypothesis has not been pursued by western medical scientists because they are convinced that particulate matter is the only plausible source of healing. However given the foregoing data, and being objective scientists, we can at least be open enough to ask whether the meridian theory has any possible theoretical foundation. We have already reviewed the accumulating data that has established the important role of bioelectric fields in fetal growth and development and have reviewed the evidence that ambient and applied electromagnetic fields can have biological effects. But we have not discussed what these fields actually are or where they come from. Since the changes reported to have resulted from acupuncture are biochemical in nature, let us begin our reflection concerning mechanisms with biochemical pathways. Many of the molecules in the body are dipoles – water molecules, protein molecules, etc. As stated previously, a dipole is a molecule sort of like a little magnet, that incorporates both positive and negative electrical charges due to an unequal distribution of its electrons. Theoretically this structure would allow the molecules to align with other dipoles in "strings," somewhat like a series of magnets. One possibility is that the energy meridians are created by dipole strings and that the structure of these strings is partially determined by the geometry of the body [4.21]. If this were the case, then the insertion of acupuncture needles might function to modulate their orientation by changing the vibration of atoms in nearby molecules, resulting in the propagation of an electromagnetic wave along the meridian. Since the chemical properties of atoms and molecules are determined by their electron configurations, which in turn determine the types of bonds they form with other atoms and molecules, a perturbation at a particular point in a meridian might be associated with a specific range of chemical changes (depending on where the needles are inserted) resulting from changes in the physical properties of the molecules and membrane potentials affected. Since traditional acupuncture theory postulates that these meridians interact, needle insertion at one point may initiate a chain reaction that stimulates other meridians and other chemical changes.

From this perspective, physical changes in energy systems would precede chemical changes of the type that have been reported in clinical acupuncture studies. This hypothesis would imply that localized chemical and physiological changes documented in acupuncture research would be one of the mediating pathways rather than the cause of clinical effects. The traditional theory states that a tumor is a localized phenomenon that spreads to the rest of the body through metastasies. The latter interpretation of events implies that tumor initiation might occur when macro-level energy meridians become blocked, causing local chemical changes that manifest at specific locations (e.g., in the form of tumors).

Chapter 2 discussed the energetic forces that are inherent in genomic and proteomic functioning: e.g., DNA bases that are dipoles, the way the negatively charged DNA strand adheres to the positively charged histone tails to prevent transcription, and the protein α-helixes which are a "macroscopic" dipoles. Covalent histone modifications (the type of chemical bond characterized by two atoms sharing an electron) play a key role in regulating DNA access for gene transcription, DNA replication and damage repair [4.22], i.e., the type of gene expression changes responsible for dysregulation or for reverting established tumor cells to normal behavior through microenvionmental cues. The basic principle involved in these bonds is electronegativity. It would take someone more knowledgeable in quantum mechanics than I, to figure out how electron structure, bond angles, dipole moments and electromagnetic spectra of simple molecules would interact mechanistically to make this happen. However loss of epithelial cell polarity has been shown to lead to increased cell proliferation and tumorigenesis [4.23-4.24], so the hypothesis that gene expression might be influenced through the perturbation of endogenous electrostatic or electromagnetic fields is not as strange as it might at first appear.

The mechanism of quantum entanglement has been proposed [4.25] to explain what happens when systemic malfunction has created cancer vulnerability by failing to repair DNA damage, or by allowing failed replication, defective telomeres, etc. As described earlier, epigenetic changes play a key role in tumor initiation and progression from benign to unstable and then metastatic. An intriguing hypothesis introduced in 2004 postulates that dipole – dipole interactions (van der Waals forces) involved in the process of creating two copies of chromosomes during cell division (mitosis) might be responsible for malignancies [4.25]. According to this hypothesis, a primary mechanism in the replication of chromosomes during the process of cell division is quantum entanglement, which causes a group of atoms or molecules to be brought into a 'quantum coherent state' (Bose-Einstein condensate), where they surrender their individual identity and behave like a quantum system governed by a wave function. Failure of quantum entanglement during mitosis would result in abnormal chromosomes leading to malignancy. This theory has enormous implications for biomedical research but has been largely ignored by the biomedical community.

4.3. DEEPER INTO THE RIFT

The scientist in me wants to stop here, in order to give the whole theory as much scientific plausibility as possible. Indeed, this is precisely what I did in my own reflections, for many years. But science is about truth, and I do not think that the theory I have just described suffices to explain either the essence of meridians or the true nature of healing. To maintain my goal of examining scientific bias, I must relate more of the personal experience that has led me to hypothesize a much deeper truth. This part is more difficult because it will probably kill any scientific credibility I might have maintained up to this point. However, the truth is that I myself have been "dabbling" in "hands on" and "psychic" healing for many years. When I was in my 20's, and my mother was ill with cancer, I went home from graduate school for two summers to take care of her. During the second of these summers, about the time I returned home, I read, "The Science of Healing," the textbook of Christian Science by Mary Baker Eddy [4.26]. I was convinced that what the author was saying could relate to the energy fields I had been hypothesizing since my earlier reading of Heisenberg, and if one just eliminated all the stuff about God, it would be quite plausible. I also knew a woman who was a Christian Scientist, who had experienced some amazing and unexplainable healings with that method. Since I did not believe in a God that picks some people to get well and others to die, I figured that if this worked, there had to be a logical explanation. I just had to figure out the mechanics of it.

The last summer of my mother's life, she was bedridden when I arrived home from graduate school. Her cancer had metastasized to the bone and she was very weak, taking heavy doses of pain medication. She knew that her cancer was entering the terminal stage. After working with her for a few weeks, doing guided meditations and other things (I was "winging" it and don't really remember what all I did, except to somehow convince her that this could be treated with "mind" and to get her to work with me though guided meditations), she was up playing golf and had greatly reduced her dosage of pain medication. This is what scientists refer to as the 'placebo effect' (believing that something will cure you, it does). I can't help wondering why, if 'belief' has the power to cause that much improvement in cancer, we aren't investigating and using it!

In September, after we had worked together all summer, I returned to graduate school. Shortly thereafter, my mother had her first set of blood tests and bone scans since before the summer. She had completely normal blood chemistries for the first time in five years and had visible healing of a number of lesions in her skeleton. Four months later she died. The September tests were interpreted by her doctors as "spontaneous remission," which is a well-known phenomenon in cancer, but is often be followed by clinical decline.

Given the resumption of an active life style, the improved blood values and healing bone lesions, she was clearly in remission. These remissions are usually only temporary and this may be the reason they have not garnered more interest or attention from the biomedical community. Yet even before this incident with my mother, I had been curious as to why oncologists did not show more interest in the phenomenon of spontaneous remission. Remission from cancer cannot happen without major physiologic and metabolic changes. What causes them and why does the patient then revert back to a dysfunctional metabolic status? If we could figure this out, wouldn't it provide useful information in our search for a cure? Is it possible that psychological changes occurring in my mother during our work together might have been responsible for physiological changes that she was not able to maintain once I had left? To an allopathic oncologist, this question doesn't even make sense, because it is assumed that causality for these diseases lies in genes, mutations, viruses, etc. Yet, as we have seen, even from the standpoint of traditional psychophysiology, there is a great deal of evidence that supports the effect of psychological states on physiological and immunological functioning.

I have related the case of my mother's cancer, as an indication of the length of time I have been wondering about these issues. At the time of her illness, I was doing post-graduate work preparing for the doctoral exams in psychology at the University of Graz, in Austria. During that period, an Austrian television station aired a German documentary about physicists from the Max Plank Institute, who had gone to the Philippines to observe "psychic" healers. These healers are usually uneducated, religiously devout people, who appear to do operations with nothing but their hands. The program was aired primarily because the wife of a prominent Viennese physician, who had been diagnosed with a back ailment that could not be cured by conventional medicine, went to the Philippines and came back cured. (Sorry, I don't remember the exact diagnosis, because it was all very peripheral to me at the time. I was preparing to become a clinical psychologist, not to do research, and I didn't take any notes.) Since her case was well documented, it created a stir in the Austrian medical community. For readers who have not heard of these healers, they stand in a large room, full of people waiting to be treated. The patient gets on the table and the healer appears to insert his or her hands into the patient's body and scoop out blood and tissue (the documentary showed actual treatments and this is the way it appeared). The tissue is then discarded. Although a film such as this can easily be faked with camera tricks or hidden tissue made suddenly to appear, the documentary had a different level of credibility. It was made by photographers accompanying physicists on an investigational visit. These physicists witnessed 100 - 200 of these 'psychic' operations. The healers treat the patients in sequence. There is no time to leave the room between individual patients in order to tuck fake blood and tissue up one's sleeve. The healers were under constant observation by trained scientists. It is difficult to imagine how it would have been possible for them to stand in a room full of people for hours on end, treating one patient after the other (probably in short sleeved shirts – the weather in the Philippines is hot) and to continue to keep pulling out fake tissue. Where they would have hidden all the slimy stuff they kept pulling out of each successive patient is difficult to fathom. What I remember most about this program (and the reason I am even relating it here) is an interview with one of the physicists. He stated carefully that he had seen "only" about 120 (I believe) of these operations and therefore did not want to generalize, but that he believed he was seeing some sort of phenomenon involving the materialization of energy. I don't remember the exact words he used, but he believed that the healers were working with energy that was then condensing into matter in their hands. What impressed me the most was that this physicist had spent quite a bit of time with these healers and believed that he was observing something real that had a rational explanation.

This film was just one more piece of information that I filed away and pondered from time to time. Another piece of information is a book by Barbara Brennen entitled, "Hands of Light" [4.27]. Barbara is a former NASA scientist, who subsequently became a healer. She claims to use energy that she channels through her own body to heal the energy fields of patients in order to bring about physical and emotional healing. She has written extensively about her methods in several books and has devoted her life to this work and to teaching it to others. But again, I read only parts of the book and put it on a shelf, still pondering.

In the meantime, I moved to Sweden, finished graduate school (having changed to a research path) and began a career in psychophysiological research at the Karolinska Institute and teaching at the University of Stockholm. During this time, my interest in energy forms of healing continued on the periphery. I married and gave birth to my daughter. At some point, I don't remember when, because I subsequently repressed all this for about 10 years, I began to practice "hands on" healing within my family. I never discussed this with anyone because I did not want to lose the scientific credibility I was beginning to establish. I was experimenting, convinced that there was an energy component as well as a physical component in human physiology and that it involved a straightforward mechanism that was only waiting to be discovered. Already at this point, I suspected that the energy component was primary. As I did these "healings," I could feel a strong flow of energy from my hands to the person I was working with, and doing this seemed to hasten the healing process of the minor everyday illnesses that I was treating. But I was "dabbling." It added another dimension to my life, it was interesting, but it was a peripheral activity.

When my daughter was five, she woke up one Sunday morning with a stomachache and couldn't walk. This was unusual for her because she almost never got stomachaches and had never had one that was painful enough to prevent her from walking. I carried her downstairs from her bedroom and laid her on the couch. She was too ill to eat breakfast. Because she was an early riser (5:30 a.m.), it was too early to call a pediatrician. As time passed, the pain began to move but she still could not eat and was not able to get up off the sofa. When it was finally late enough to reach a doctor, I did. Because a stomachache was not considered an emergency, I could not get an appointment with a pediatrician before 2:00 p.m. By 10:00 a.m. (about 4.5 hours after she had awakened), she was still in pain and I became concerned because she had never experienced anything like this before. I started doing some "hands on" healing because there were still four hours to go until the 2:00 p.m. appointment. As I began, I suddenly moved into a different state of consciousness. It was a very deep sense of love. After doing the healing for a short time (10 - 15 minutes?), I stopped. About a half hour later she suddenly jumped off the couch and began to play as if nothing were wrong. She ate normally and seemed completely healthy. Despite this, I kept the afternoon appointment with the pediatrician. When I described the symptoms to the pediatrician, she said that it was a classic case of appendicitis and said she could not understand why my daughter was no longer showing any of the symptoms. Needless to say, I just shrugged my shoulders and said nothing.

One would think that this would have been a very positive experience for me. But in fact, it was very disturbing. I remembered how much more powerfully I had felt the energy flow from my hands when my state of consciousness shifted. It was different than anything I had experienced before and the implications were very upsetting to me. This was not what I had bargained for. I had been convinced that what I was experimenting with was an objective physical phenomenon that could be treated in a "mechanical" way. Something you could, in principle, explain with an equation. It was simply a transfer of energy and if I could just figure out the mechanism, it would fit into traditional biomedical research. Now I was faced with the possibility that my state of consciousness could influence the energy flow and I wanted no part of it. It was too subjective and totally unacceptable to my scientific frame of reference. I may have begun to believe in God, but as far as I was concerned, I didn't need God to do science and even the suggestion that my subjective state of mind could influence the energy flow, "freaked me out."

What I did, was to go into a state of denial. I decided that what these events indicated was that I didn't have a clue what I was doing, and that even though I was still convinced that energy healing was possible, it was best left to people who understood it. I certainly did not, and I made a decision to stop doing the healing altogether. In the future I would simply stick to research. I hoped I would someday find a way to investigate these issues, but purely as an "objective" observer. What I could not deal with psychologically, was the powerful energy flow that I had felt when I experienced intense feelings of love. It was just too subjective. I wanted a simple mechanical principle that could be explained without a lot of subjective "mumbo-jumbo." I "forgot" about my daughter's healing and about my mother's and went about my business doing mainstream stress research. For ten years I went through the motions, but became less and less enthusiastic about the traditional research I was doing.

I was not investigating what seemed to me to be the central issues, namely the biophysics of energy healing, and I was frustrated that there was so little opportunity for the type of research I really wanted to do. The questions that interested me were not relevant to mainstream science and could not be funded through traditional research resources.

I had reached a point of complete frustration. I felt that the science I was doing was just one more permutation of the same paradigm and I couldn't see a way to fund the research I really wanted to do. One morning, although I have never been very good at meditating, I decided to meditate. All of a sudden, during the meditation, I remembered my daughter's healing and how upset I had been at the element of subjectivity that had entered into it. I remembered that ever since then, when talking to the few people I knew who were interested in the topic I had always stated flatly that I wasn't a healer but I was interested in the phenomenon. I realized that I had been splitting off a part of myself and that from a psychological perspective this type of fragmentation was not healthy. The part of me that had been "split off" needed to be integrated into my current life. But again, I was so skeptical and full of doubt that I didn't really know where to begin. However, I did make a conscious decision to face up to the problem that had caused me to stop the healing, although I was (and still am) very skeptical of doing any healing myself.

A couple of months after that meditation, I "remembered" the spontaneous remission my mother had experienced after I had worked with her. Almost a year later I also "remembered" that I had actually started experimenting with "hands on" healing as a child. I must have been about 9 years old when it first happened. I had been born with a mole on the front of my leg, directly under the knee, that had begun to change color, become uneven and seemed to be growing. Even at that age, I had heard about the indications of cancerous moles and decided to see if I could 'heal' the mole with my hands. In Sunday school I had been taught that Jesus healed with his hands and that he had said, "All this you can do and more." and I believed it. Every day I held my hand over the mole for a few minutes and at the end of two weeks the mole that I had been born with had completely disappeared. I remember being rather pleased with myself but not thinking much more about it. No one else noticed.

Psychologically, I believe that my decision to stop splitting off the part of myself that was involved with healing, led to my remembering events that I had repressed. The relevance of this part of my personal journey for the points being made in this book is that it reveals some of my personal bias in hypothesizing relevant mechanisms. None of the anecdotal incidents I have related is "proof" of anything. But they are part of the sum total of experience that contributed to the formation of my hypotheses. These experiences, together with the data in the previous chapters and the information in the next chapter on consciousness, are part of the experience that has lead me to question some of the axiomatic tenets of biomedical science and to hypothesize a different concept of health and illness than the traditional one. They have led me to begin to question whether consciousness was somehow involved with energy healing.

This question is extremely important to the issue of how scientific bias influences the methods that we use in designing experiments. In fact, it goes to the heart of the issue. Now that the Center for Complementary and Alternative Medicine at the National Institutes of Health has made funds available to research alternative treatments, scientists interested in the area have struggled hard to come up with objective methods to investigate them. They have tried as much as possible to use the models created by traditional medicine in their conduct of research. Since the basis of these models is the ability of the person treating the patient to remain aloof from the treatment (e.g., there is no involvement of consciousness), I would like to briefly describe the traditional methodology and indicate how the belief system it utilizes may be influencing the results.

One of the primary objectives in traditional clinical trials is to control not only for unintentional bias, but also for the "placebo" effect. As mentioned above, the placebo effect involves people getting better, not because of the active drug they are taking, but because they believe they will. I could go off on a tangent about how important it is to actually investigate the placebo effect instead of controlling for it, but I will temporarily refrain. What we do to test for "assay sensitivity," which is another way of saying "drug effectiveness," is to randomly assign people with the same disease to two groups – one with a real drug and one with a "placebo." A placebo is an inert substance, like sugar, that looks like the real thing. Where possible, the doctors are also blinded as to which patients are getting the active drug. When both doctors and patients are unaware of who is receiving the active treatment, it is called a "double blind" experiment and is considered to be the ideal approach. The reason for this methodology is that we want to make sure that the doctors treat all the patients in exactly the same manner and do not favor the ones getting the drug. Based on these methods, we assume that the results we get from these trials are completely objective. Of course, one of the problems with blinding is that it often doesn't work because the group getting the real drug exhibits side effects that are not present in the control (placebo) group. So, both doctors and patients sometimes have a good idea who is actually getting the active drug. Nevertheless, it is the best we have been able to do in order to maintain objectivity.

In cancer trials, however, there usually is not a placebo arm. For ethical reasons, patients are not assigned to placebo but to other treatments. The control group may be assigned to a treatment that is currently in use and the experimental group, to one that is new. There is always a "Data Safety and Monitoring Board" (DSMB) that oversees the trial and monitors the results. All "adverse events" must be reported. Once a critical number of patients has received the drug, if one drug is clearly more effective or causing more adverse events, the trial may be stopped. Every attempt is made to insure patient safety and to assure that what the patients and doctors think and feel does not influence the results.

In order to achieve objectivity in "alternative medicine", scientists have also gone to great lengths to eliminate any subjectivity. A good example is the use of energy healing on cell cultures [4.28], developed by Garret Yount. To avoid the problem of belief systems, he has adapted standard experimental protocols from molecular oncology to test whether Qi gong (a form of energy healing) directed at brain tumor cells can accelerate their death and whether Qi gong can accelerate the growth of normal cells. The Qi gong practitioners sit on a laboratory bench and are instructed to maintain "neutral intentions." This design attempts to keep the issue to an "objective" transfer of energy. And in 35 trials with appropriate controls, they have found a small but significant difference in cell growth between the treated and untreated cells. The problem is that this methodology skirts the issue of the mind. First of all, this is not what healers do when they heal. They have clear intentions of directing healing energy toward the patient. Second, if my own experience with my daughter is any indication, the subjective state of the healer may be an important component in the healing process.

Why is this important? We know that the neurons in the brain are associated with electrical impulses and that we can measure brain wave frequencies on the surface of the scalp. Actively thinking is associated with different frequencies than relaxation. When a person's attention shifts, this can also be measured as a change in the shape of a particular type of brain wave called an evoked potential. If thoughts are associated with wave frequencies and these frequencies can be measured on the outside of the brain (EEG), then they are also part of what physicists refer to as a force field. If there is such a thing as healing based on changes in physical force fields, (energy), then active emotional involvement and a "neutral" attitude may have be associated with different frequencies and have entirely different effects. This means that attitude would be a variable to be investigated, not controlled for.

At this point it may also be appropriate to point out that for most people involved in energy healing, there is a belief that we are spiritual beings having a physical experience and that we are not meant to stay in this dimension forever. The "healing" may involve healing the body, healing the mind, or easing the struggle of a terminally ill patient by making the transition (dying) easier. The mind is very much present in this work. The ability of the mind to promote or inhibit healing through the strength of its beliefs, may be one of the most powerful effects there is and should be a focused target of investigation.

Had I not faced up to the implications of my past experience healing my daughter and begun to acknowledge the role of consciousness in the healing process, I would still be trying to eliminate the mind from my hypotheses. The deleterious effect that personal bias can have on science will be made more evident in the next chapter.

REFERENCES

4.1) Wong AM, Su TY, Tang FT, Cheng PT, Liaw MY. Clinical trial of electrical acupuncture on hemiplegic stroke patients. *Am J Phys Med Rehabil* 1999;78:117-122.

4.2) Hollinger I, Richter JA, Pongratz W, Baum M. Acupuncture anesthesia for open heart surgery: a report of 800 cases. *Am J Chin Med*;7:77-90.

4.3) Aune, A., Alraek T, LiHua H, Baerheim A. Acupuncture in the prophylaxis of recurrent lower urinary tract infection in adult women. *Scand J Prim Health Care* 1998;16:37-39.

4.4) Aune A, Alraek T, Huo L, Baerheim A. Can acupuncture prevent cystitis in women? *J Norweig Med Assc* 1998;118:1370-1372.

4.5) Richter A, Herlitz J, and Hjalmarson A. Effect of acupuncture in patients with angina pectoris. *Eur Heart J* 1991;12:175-178.

4.6) Hu J. Clinical observation on 25 cases of hormone dependent bronchial asthma treated by acupuncture. *J Tradit Chin Med* 1998;18:27-30.

4.7) Timofeev MF. Effects of acupuncture and an agonist of opiate receptors on heroin dependent patients. *Am J Chin Med* 1999;27:143-148.

4.8) Wu B, Zhou RX and Zhou MS. Effect of acupuncture on immunomodulation in patients with malignant tumors. *Zhonggup Zhong Xi Yi Jie He Za Zhi* 1994;14:537-539.

4.9) Wu B, Zhou RX, Zhou MS. Effect of acupuncture on interleukin-2 level and NK cell immunoactivity of peripheral blood of malignant tumor patients. *Chung Kuo Chung Hsi I Chieh Ho Tsa Chih* 1994; 14:537-539.

4.10) Yuan J and Zhou R. Effect of acupuncture on T-lymphocyte and its subsets from the peripheral blood of patients with malignant neoplasms. *Chen Tzu Yen Chiu* 1993;18:174-177.

4.11) Ahu Dan and Tian Ming. Changes of visual findings, electric features and staining of auricles in malignant tumor patients. *J Tradit Chin Med* 1996;16:247-251.

4.12) National Institutes of Health Consensus Development Conference on Acupuncture. *Acupuncture 1997;15:1-5.*

4.13) Kim E, Kim Y,Jang M, Lim B, Kim Y, Chung J, Kim C. Auricular acupuncture decreases neuropeptide Y expression in the hypothalamus of food-deprived Sprague-Dawley rats. *Neurosci Lett* 2001;307:113-6.

4.14) Chen BY. Acupuncture normalizes dysfunction of hypothalamic-pituitary-ovarian axis. *Acupunct Electro-Ther Res* 1997; 22:97-108.

4.15) Tsuei JJ. The science of acupuncture - theory and practice. *IEEE Eng Med Biol Mag* 1996;15:52-57.

4.16) Shang C. Mechanism of acupuncture –beyond neurohumoral theory. *Med Acupuncture Online J* 1999/2000; 11(2).

4.17) Ellsworth DL, Ellsworth RE, Love B, Deyarmin B, Lubert SM, Mittal V, Hooke JA, Shriver CD. Outer breast quadrants demonstrate increased levels of genomic instability. *Ann Surg Oncol* 2004;11:861-868.

4.18) Kolberg R. Angiogenic inhibitor loss may be key to post-surgery metastasis. *J NIH Res* 1994;6:31-33.

4.19) O'Reilly MS, Holmgren L, Shing Y, Chen C, R, Rosenthal RA, Moses M, Lane WS, Cao Y, Sage EH, Folkman J. Angiostatin: a novel angiogenesis inhibitor that mediates the suppression of metastases by a Lewis lung carcinoma. *Cell* 1994;79:185-188.

4.20) Heisenberg W. Physik und Philosophie. Weltperspektiven Band 2. Deutche Originalausgabe. Ullsten Bücher. Germany 1959.

4.21) Knox SS. Physics, biology and acupuncture: Exploring the interface. *Frontier Perspectives* 2000.9:12-17.

4.22) Chi P, Alls CD, Wang GG. Covalent histone modifications- miswritten, misinterpreted and mis-erased in human cancers. *Nat Rev Cancer* 2010;10:457-469.

4.23) Bissell MJ & Radisky D. Putting tumours in context. *Nat Rev Cancer* 2001;1:46-54.

4.24) Gudjonsson T, Ronnov-Jessen L, Villadsen R, Rank F, Bissel MJ, Ptersen OW. Normal and tumor-derived myoepithelial cells differ in their ability to interact with luminal breast epithelial cells for polarity and basement membrane deposition. *J Cell Science* 2002;115:39-50.

4.25) Hameroff SR. A new theory of the origin of cancer: quantum coherent entanglement, centrioles, mitosis, and differentiation. *BioSystems* 2004;77:119-136.

4.26) Eddy MB. *Science and health with a key to the scriptures.* The First Church of Christ Scientist, Boston, Mass. 1903.

4.27) Brennan B. *Hands of Light.* Bantum Books, New York, 1987.

4.28) Schlitz M, Lewis N. Subtle realms of healing. *IONS* 2001;55:30-37

CHAPTER 5

♦

CONSCIOUSNESS

After almost 50 years of neurosurgery to treat epilepsy, the neurosurgeon, Wilder Penfield made the following statements:

> Electrical stimulation can cause the patient…to turn head and eyes, or to move the limbs, or to vocalize and swallow. It may recall vivid re-experience of the past, or present to him an illusion that present experience is familiar. But he remains aloof… There is no place in the cerebral cortex where electrical stimulation will cause a patient to believe or decide. …there are also, areas of gray matter in the higher brain-stem that the surgeon's stimulating electrode does not explore. Yet epileptic discharge can take place in any area of gray matter… There is *no* area of gray matter, as far as my experience goes, in which local epileptic discharge brings to pass what could be called "mind-action" …For my own part, after years of striving to explain the mind on the basis of brain-action alone, I have come to the conclusion that it is simpler (and far easier to be logical) if one adopts the hypothesis that our being does consist of two fundamental elements… [5.1]

He states further, "Meantime we can only use the language of dualism, and speak thus for the mind and the brain… The ancient riddle of how brain and mind interact is still unresolved[5.2]."

The question of whether or not brain (matter) gives rise to mind, or mind interacts with brain, has long been central to the discussion of human consciousness. Many large, academic tomes have been written on this topic and an overview would be beyond the scope of this book. The intent with this chapter is to indicate the complexity of the issue by illustrating how approaches to the study of consciousness differ, depending on the investigator's beliefs.

5.1. PSYCHOPHYSIOLOGY AND NEUROPSYCHOLOGY

Since much of the debate concerning conscious processes centers around the physiology of the brain, a good place to begin the discussion is with an overview of two of the medical disciplines that investigate how the brain interacts with the body, namely, psychophysiology and neuropsychology. The field of psychophysiology focuses on how psychological states such as anger, distress, and love influence the physiology of the brain and body. Neuropsychology investigates how organic changes or anomalies in the brain affect cognitive and psychological functioning. To put it simply, the first of these investigates how "mind" influences matter, and the second investigates the influence of matter on "mind." Both of these disciplines have produced abundant, well-documented data.

Decades of psychophysiological research in animals and humans has demonstrated that what we think, feel and experience can have profound effects on the physiology of both the brain and the body. Beginning with lower vertebrates, a series of experiments performed on fish has demonstrated that manipulation of social status, such that non-territorial fish become territorial (dominant) and territorial fish become non-territorial (submissive), results in changes in growth rate [5.3] and the size of growth related neurons in parts of the brain [5.3]; as well as changes in a stress hormone (cortisol) and in the ability to reproduce [5.4] due to size changes of hormone-containing neurons in the forebrain [5.5]. It also causes changes in brainstem concentrations of a substance (serotonin) [5.6] known to be associated with depression in humans. This research concerning the influence of dominance status on multiple physiological systems has demonstrated that during development, behavior interacts with the environment to cause structural changes in the brain that last a lifetime, and that social behavior in adult animals can cause permanent changes in brain functioning [5.7].

Research on higher vertebrates, such as non-human primates, whose physiology is very similar to that of humans, reveals even more evidence of the profound effects that psychological experience can have on the physiology in the brain and body. The reason that this research is so important is that the psychosocial environment can be experimentally manipulated in a way that mimics real life stress but under conditions that control for all extraneous factors that might influence the results (e.g., diet). It has been demonstrated that the social manipulation of monkeys, resulting in subordination in some and dominance in others, similar to the fish experiments above, produces a depressive response in the subordinates that is accompanied by increases in cortisol secretion (a stress hormone) and suppressed reproductive function [5.8].

Subordinate monkeys also develop dysfunction in the arteries of the heart that is not found in dominant monkeys [5.9]. The administration of a substance (acetylcholine) that in healthy monkeys causes the arteries to expand (dilate) has the opposite reaction in the subordinate monkeys. The arteries constrict, causing the heart to receive less oxygen. Lack of oxygen is what causes heart attacks and both blockages and narrowing of the arteries can contribute to this. Not surprisingly, subordinate monkeys also develop more blockage (coronary artery atherosclerosis) than dominants when both are fed exactly the same diet [5.10]. In addition, like premenopausal human women, female macaque monkeys are more resistant to atherosclerosis than males. However, this protection is lost in subordinate females, who resemble males with respect to the amount of plaque in their arteries [5.11]. Together, this group of experiments demonstrates that psychological stress can cause dramatic physiological changes in both the brain and body.

One of the most well-known examples of the opposite effect, namely how physiologic changes in the brain can influence psychological processes, comes from the neuropsychological literature on the chronic degenerative brain disease, Alzheimer's. In addition to genetic contributions, toxicants in the environment [5.12] and hyperthyroidism [5.13] have been postulated to contribute to Alzheimer's, although the etiology is still unclear. This type of dementia is so prevalent that almost everyone has had an acquaintance, friend or extended family member who has struggled through the memory loss and negative personality changes that develop as the disease takes its toll, while the victim sinks deeper and deeper into what appears to be cognitive oblivion. Alzheimer's disease is an extreme example but there are many other, more subtle examples of how organic changes in the brain can influence behavior. Two such examples are the lack of oxygen at birth (hypoxia) on behavioral response to stress in adult rats [5.14]; and cognitive deficits caused by multiple sclerosis [5.15]. But even more common examples include such every day experiences as how our mood changes when we are in pain or when we drink alcohol.

5.2. PSYCHIATRY AND THE MEDICAL MODEL

What is extremely clear from the array of data accumulating in the fields of neuroscience and psychophysiology, is that the interaction between the brain and what we feel, think, and experience, is highly complex. Since theories postulating that all causality can be traced to matter are consistent with only a small part of this data, and directly contradict other aspects, they are too simplistic to account for much of what we know about human consciousness. These facts notwithstanding, psychiatry, as opposed to psychology, is a medical subspecialty, firmly anchored in the biomedical frame of reference. The predominant belief in the field of psychiatry coincides, therefore, with the materialist frame of reference in medicine. It assumes that all aspects of the human mind, including feelings, cognitions, self-reflection, and transcendent states, originate in and can be traced to the biochemistry of the brain, possibly associated with genes. States of psychological discomfort or disturbance, be they depression or schizophrenic hallucinations, are all lumped together under the umbrella of "mental illness" and most are believed to be the result of bad genes or chemical imbalance. This belief in the biochemical or genetic causation for all of human experience is called the "medical model" of psychiatry. It follows from this model, that since the cause is biochemical, the cure should also be sought in biochemistry, i.e., in pharmaceutical products. As the above data indicate, there are a number of factual errors in this theory. The first is the belief that knowing the concentration of a particular biochemical substrate such as a neurohormone or neurotransmitter associated with a particular condition, tells us something about causality. In order to understand causal relationships, we need to know more about the context in which a biochemical change occurs. The second factual error concerns the inaccuracy of the materialist assumptions on which the medical model is based. This has already been discussed and will be illustrated by concrete examples of its maladaptive application with respect to treatment and interpretation of data.

5.2.1. Depression

A major example of the consequences that erroneous reasoning can have in clinical practice is the treatment of depression, a debilitating condition which is more common in women than in men [5.16]. Biochemical studies, although not totally consistent [5.17] have repeatedly reported abnormalities in serotonin and norepinephrine receptor function in the brains of depressed patients [5.18-5.21]. Both of these chemicals are neurotransmitters, which means that they help to transmit a signal across a synapse, or gap, between nerve cells. Given the existing data from psychophysiology and neuropsychology, the obvious question for an objective scientist is whether this dysfunction is related to the causality of depression, is a mediator (i.e., lies somewhere between the cause and the effect), or results from it. The answer is that we simply don't know without more information. What we do know from animal research, is that serotonin metabolism can be influenced by stress [5.22-5.25] and that the expression (function) of the serotonin and other genes interact with and can be influenced by the environment (e.g., early rearing conditions). This means that the gene is expressed one way under certain environmental conditions and another way under others [5.26-5.27]. However, these data on the intimate interactions between genes, environment and biochemistry, don't seem to have had any major influence on the most commonly prescribed therapeutic regimens, namely pharmaceuticals targeting serotonin levels.

Psychiatrists routinely make a logical "leap of faith" when they interpret the data on altered serotonin or cortisol function in depression as an indication of biochemical causality. Although the presence of psychological contributors would be acknowledged by many psychiatrists, they are considered adjunct to the basic biochemical abnormality and are rarely taken into consideration when prescribing treatment. Logically, this jump from association to causation is analogous to saying that the wheels of a car are the cause of its acceleration up a hill because they move when the car moves, increase the frequency of their rotation when the car speeds up, and stop when the car stops. In scientific terms, wheels are what we call a "marker" or indicator, something associated with acceleration, but not causing it. What causes acceleration is the motor. A car with wheels but no motor would roll downhill, not accelerate up a hill. Similarly, darkness may be associated with seeing the moon but the moon is not causing the darkness. Rotation of the earth away from the sun causes darkness. A correlation between neurotransmitter dysfunction and depression tells us nothing about causality.

Despite this rather obvious fact, pharmaceutical products targeting neurotransmitter function have become the standard model for treatment of depression. If one conceptualizes that a possible genetic susceptibility, i.e. a specific polymorphism of the serotonin gene, has interacted with environmental conditions to result in a manifestation of depression, then logic would indicate that a combination of biochemical and psychological treatment modalities might be the most effective way to begin treatment because multiple contributors can be targeted simultaneously. Unfortunately, a serious psychological treatment plan (other than short, periodic supportive meetings with the psychiatrist) is not part of the standard treatment regimen and is difficult to cover with most health insurance plans. The effectiveness of antidepressants has been called into question by a meta-analysis of multiple studies. It would seem that their benefit is not much better than placebo in mild and moderately depressed patients but helps in severely depressed patients [5.28].

What this indicates is that demonstrating the existence of a biological correlate to depression does not give us enough information to understand the causal mechanisms or to plan the most effective treatment regimen. Do the fallacies associated with the medical model mean that drugs should not be used to treat depression? No, of course not. Although clinical trial data on selective serotonin reuptake inhibitors are contradictory, the medications seem to reduce symptoms in some people. The issues is that we need to know more about how the receptor dysfunction fits into the context of depression – whether it contributes causally, is simply a marker that covaries with some other causal factors, or is a consequence of depression. Since animal research has demonstrated unmistakably that changes in serotonergic metabolism can result from stress [5.22, 5.24-5.25] and human research has demonstrated that serotonergic genotypes are associated with vulnerability to depression – but only in adverse environments [5.29], before drawing conclusions about causality, we need more information.

However, even if neurotransmitter function plays a mediating rather than a causal role, pharmaceutical treatment of that part of the pathway in conjunction with other treatments may be beneficial. The reason we need to know about causality is so that we can formulate a more effective treatment strategy that targets all of the relevant components. It is important to point out that, unlike antibiotics used to treat infection, which are taken for a limited number of days until the cure is complete, anti-depressants do not cure depression. They eliminate some of the symptoms in some of the patients, part of the time. This is good as far as it goes, but not enough for us to sit back and assume that we have discovered the cause and cure of depression. Medication for depression, to the extent that it does alleviate symptoms, works like aspirin taken for pain relief. When you stop taking it, the symptoms return. The efficacy of antidepressants in comparison with placebo is not well established [5.30] and many of the common beliefs about current anti-depressant medications are not adequately supported by the scientific data [5.31]. We still need more research to improve treatment efficacy.

Most of us are aware that when we get upset, our muscles tense, our stomach is "in a knot," our heart starts beating faster, or we might begin to perspire. Even though these physiological changes are a result rather than a cause of the stress, taking a tranquilizer will have a biologically calming effect and because we are calmed we will probably feel better. Are tranquilizers the best treatment strategy for stress? Some people would probably answer "yes" to this question, others "no". Still others might reply that the most effective long-term strategy involves investigating and removing the cause of the stress, and that if taking tranquilizers while the stress is being eliminated is helpful, so much the better.

Translated to depression, what might be some of the potential contributors to causality? An animal model developed in macaque monkeys has contributed some tantalizing clues. The model, created by Suomi et al. [5.32], involves the rearing of infant monkeys in peer groups of four without any mothers. When subjected to repeated separation from their peer group, the peer reared monkeys first reacted hyperactively but subsequently began to exhibit depressive behaviors such as self-clasping, huddling, passivity, and social withdrawal. Suomi's group then showed that these monkeys could be rehabilitated to behave in an age appropriate socially adaptive manner, by being repeatedly exposed to "therapist" monkeys. Therapist monkeys were selected on the basis of their socially active stimulus behavior. The therapy consisted essentially of planned interactions between the monkeys, wherein the therapy-monkeys, based on their natural predisposition, enticed the depressed monkeys out of their withdrawal [5.33]. In a summary of work on animal models of depression [5.34], Suomi points out how difficult it has been to trace the effect of early life experience on the development of subsequent depression in humans. The reason is that retrospective data, as abundant as they are in the clinical literature, do not allow one to draw conclusions about causality, because of the plethora of intervening variables that might also have influenced the outcome. However, a review of the animal research clearly demonstrates that inadequate early social attachment experiences can have long-term biobehavioral consequences for macaque monkeys. Despite the caveats, similar findings have been reported in human research where people experiencing early parental loss have been shown to have significantly increased odds of developing major depression in adulthood [5.35]. What we also know from prospective human research is that there are important interactions between genes and environment indicating that individuals with one or two copies of what is called a "short" allele of the serotonin transporter gene have been shown to react with more depression to stressful life events than those that have two "long" alleles of that gene [5.29].

A long series of experiments in non-human primates has revealed some very important information about interactions between heredity and environment that seem highly relevant for the understanding of human depression [5.36]. In the wild, about 20% of most populations of rhesus monkeys seem to be born with enhanced physiological arousal causing them to be highly reactive to novel stimuli. They display anxiety and increased physiological arousal, including cortisol secretion and the noradrenergic activity discussed above in the medical model of depression. When young, these monkeys often lapse into behavioral depression during times of temporary maternal absence during the annual breeding season. Thus, like the "short form" genotype mentioned above, they seem to be genetically susceptible to stress. This anxious behavior is also accompanied by greater physiological reactivity in certain parts of the brain (the hypothalamic-pituitary axis) than in normal monkeys. Surprisingly, cross-fostering studies show that these tendencies can be permanently overcome by placing such infants, with highly nurturing foster mothers for the first six months of life. In fact, the high anxious infants placed with nurturing foster mothers not only lose their high-reactive behavior, but became behaviorally precocious, progressing ahead of what would normally be expected from their chronological age. When these cross-fostered high-reactive infants were removed from their foster mothers and placed in larger social groups, subsequent follow-up revealed that their social adjustment had continued and that they were highly adept at social alliances [5.34].

The above data indicate that there may be multiple contributors to depression, including a genetic vulnerability and negative early childhood experience. It is also a hopeful model, because the importance of early childhood experience means that the affected monkeys could be cured through "therapeutic" treatment. The curative effect of "therapy" monkeys in this research might indicate a constructive role for psychotherapy in human depression, as well. But this does not fit the medical model so qualified therapists are difficult to obtain through health insurance plans.

Another result of the overly simplified emphasis on biochemistry in the psychiatric community are the pharmaceutical ads on television, that have educated the general public to believe that if they feel sad or 'bad', the best solution is to take a drug. This situation creates an interesting contradiction in the message we give young people. On the one hand, they see their parents and other adult role models reach for drugs to improve their mood, on the other hand, the adults tell young people who are in the painful turmoil of adolescence, that "taking drugs" is not a way to solve problems or create lasting happiness. This is a very ambivalent message and given the statistics on drug abuse, one that is clearly not working.

The foregoing data have highlighted three possible contributors to depression: biochemical changes, genetic susceptibility and early childhood experience. There is also another set of data that might provide additional insight, namely data on electroshock therapy, which has been shown to give temporary symptomatic relief for some very depressed patients who cannot be helped with drugs. The changes achieved can be dramatic but the relief achieved with electroshock is usually short lived and often demands a high price in the form of confusion and memory loss [5.37-5.44] which seem to accumulate with each subsequent treatment. Although it results in great temporary relief in some deeply depressed people, no one knows what the mechanism is. I believe that referring back to nonlinear dynamical theory (Chapter 2) might help to elucidate this. The cognitions of people who are psychologically depressed reflect a loss of flexibility. Their thinking becomes "mode locked" in thoughts of hopelessness and they see "no way out". Hopelessness is their psychological "phase space." just as the mass of taffy was the phase space for the food color that was being swirled into it (see Chapter 1). In the same way that coloring could be distributed to different places within the taffy but could not go outside of it, the thoughts of a depressed person do not seem to be able to leave the sphere of negativity and hopelessness. This state of affairs is very similar to what in mathematical terms is called a "single attractor." An attractor is a state of relative stability into which nearby solutions settle [5.45]. A single attractor means that there is only one steady state, rather than multiple states. Psychologically, the stable state of hopelessness in depression does not allow the healthy flexibility required to meet the normal challenges of everyday life.

In the discussion of complex systems, it was pointed out that a healthy system needs flexibility to function properly and respond to the varying demands being made upon it, and that when this flexibility is reduced to rigid patterns it is an indication of ill health. On the basis of what we know about dynamical systems, it could be speculated that along with their cognitions, the brain waves (EEG coherences) of depressed people have also become mode locked. What might be needed psychologically for people with depression, is to somehow perturb the system just enough to knock it away from a single attractor and get it back into a state of healthy variability. A chaotic, healthy, system often exhibits what is called a "strange" attractor that is neither a steady state, nor periodic, but has a limited range and is stable enough so that it is not destroyed by small changes. If it is the case that both the cognitions and brain frequencies of depressed people are "mode locked," there might be several ways to perturb the system. Psychologically, this might be accomplished with psychotherapy. Electrophysiologically, a small electric stimulus strategically placed, might break up the electrophysiological single attractor and perturb the brain waves into greater variability, at least temporarily. If this is the case, it might explain the mechanism for the temporary relief provided to some patients by electroshock (ECT) treatment. However, it would also imply that the currently available techniques, which instigate violent brain seizures, would be the neurological equivalent of an elephant in a porcelain shop. It would indicate that what is needed is the use of nonlinear dynamics to calculate a smaller, more directed perturbation to achieve the same symptomatic relief, but without the memory loss. A more targeted treatment might possibly also be more effective. Since the amount of memory loss with the current method increases with the degree of symptom relief (5.43), nonlinear solutions to this problem would be a worthwhile area of research to pursue. However, both of these solutions (perturbation through psychotherapy or through ECT) beg the question of causality. We still don't know which comes first, the psychological single attractor or the physiological one. What we have seen is that the positive psychological effects achieved by electroshock are only temporary, indicating that we need more information. Does this indicate that cognitions are causing the electrophysiological patterns to return to a single attractor and that future dynamical methods should be used in conjunction with psychotherapy to prevent relapse? This is only one of several issues that we could and should be addressing if we are to improve treatment efficacy.

If we step back and review the data on depression, it becomes clear that biochemical involvement is only one aspect of a complex system. That the field of psychiatry has chosen to focus almost exclusively on this aspect, to the exclusion of other treatment modalities, is a result of a belief that causality is reducible to the particulate characteristics of matter. There is a natural limit to the number of variables that can be included in any one experiment and scientists base their choice of variables on what they believe to be relevant. The problem is that the important role of belief in these choices is not acknowledged. Both the scientific community and the patients it treats would greatly benefit from a discussion of this issue. For if a belief system determines what variables should be included in the investigation then an incorrect belief system may mean that important variables are being ignored.

5.2.2. Near Death Experiences

This is nowhere more evident than in the study of "near death" (ND) experiences. People who have had close experiences with death sometimes relate experiences which share common characteristics. They may have experienced being in a tunnel and being drawn toward a bright light, feeling surrounded by love, experiencing a Divine Presence, or receiving messages from "heavenly beings." Sometimes these beings tell them that it is not yet time for them and that they must go back. Accumulating observations documenting these experiences, not available even two decades ago, and published by reputable investigators [5.46], have not lead any scientific discipline, including psychiatry or psychoanalysis, to open a line of inquiry into whether there is life after death, or to ask what "near death" experiences imply about the nature of human reality and the existence of God. The scientific responses have been primarily limited to explaining away these experiences as drug (e.g., ketamine) [5.47-5.48] or hypoxia (lack of oxygen to the brain) induced hallucinations, or mental illness [5.49-5.52]. One well-documented aspect of these experiences in people who have died during an operation and been subsequently resuscitated, is that they claim to have risen above their bodies and looked down on what was taking place in the operating room while they were clinically 'dead'. These people have accurately recounted exactly what went on during this period when they were unconscious or their hearts had stopped beating. However, these facts have not given this issue enough legitimacy in scientific circles to be taken seriously as a basis for investigation of these phenomena. This part of the data is simply ignored and the experiences themselves are treated as epiphenomena, symptomatic of altered physiological states or underlying psychopathology. Accounts of having encountered a Divine Presence or loving beings who communicated with them, makes the teller unbelievable in the eyes of scientists, because in the current scientific frame of reference, such beings do not exist. Therefore, anyone reporting having seen them must be hallucinating. This is, of course, is circular reasoning. It amounts to saying, "Since these data contradict my beliefs about the nature of reality, they must be inaccurate and should therefore be ignored." The crucial issue is that the driving force in the reasoning of these scientists is not the data itself, but their personal beliefs about what the data mean. Just as the papal doctrine formulated the frame of reference for science during Galileo's time, the materialist doctrine is dictating it today. *Despite the scientific claim of objectivity, beliefs, not data are driving science.*

What is curious about the reaction of the scientific community to 'near death' experiences is the general lack of interest in a phenomenon that could have such profound implications for the nature of reality and human existence. If what these people from all socioeconomic backgrounds, cultures and belief systems are experiencing is real, does this not have profound implications for us as human beings? Is the issue of life after death and what it implies not one of the most important scientific questions we could ask?

Being human, we all have biases. As scientists, our frame of reference, i.e., our beliefs about the fundamental nature of reality influence where we look for answers to specific scientific questions and the variables we include in our experiments. They even influence what is defined as science and the questions that constitute legitimate areas of scientific investigation. Since no single experiment or series of experiments can ever include all possible variables, selection bias is an unavoidable, constant presence in scientific research. The problem is not that scientists have biases, but that they are largely unaware of them and thus, risk propagating them as objective fact. Worse, our biases limit our methodological approaches to the investigation of scientific questions, as well as the definition of what constitutes a scientifically relevant topic. Scientists define what science is and scientists have decided that the existence or non-existence of God is not a scientific question. In the example of "near-death" (ND) experiences, a materialist (someone who believes that physical matter is the only fundamental reality and that all causality is a manifestation of a matter substrate) will look for the explanation to the experience in the chemistry of the brain or might hypothesize hallucinations associated with psychopathological states. These are valid hypotheses. What is problematic is that these scientists do not acknowledge even to themselves that there are legitimate, non-materialist hypotheses that also fit the existing data. The possibility that people who claim to have experienced a Divine presence may have actually done so (without being mentally ill or having hallucinations) is not even on the table for discussion. This is not objective science.

The scientific bias that there is no God and therefore no after-life means that these possibilities are not eligible to be investigated as explanations for ND experiences. For some reason, *the issue of Divinity is taboo* and any scientist who openly broaches it, risks jeopardizing his or her career. It is as if we have come full circle from the time of Galileo. During that period in history, any theory that implied that God might not be the center of the universe was heresy. For describing planetary motion and defending Copernicus' heliocentric view, Galileo was tried by the inquisition, forced to recant, and kept in house arrest for the remainder of his life. Today, it is heresy to broach the opposite as being scientifically relevant, namely that God might actually exist. However, a scientist who does so, no longer risks arrest or trial, s/he risks instead being marginalized as a scientist – losing scientific credibility and facing a great deal more scrutiny in peer review for scientific journal articles and grant applications. Psychoanalysts who deviate from the materialist paradigm risk something else -- losing patient referrals. In the fields of psychiatry and biomedical research, the non-existence of God is so basic an assumption that it is considered axiomatic. The amazing part of this is that this axiom is not based on any scientific evidence. There is no established scientific proof to support it!

Interestingly, the field of psychoanalysis, which is intent upon investigating and demonstrating the existence of the unconscious and the importance of psychological phenomena such as defenses, drives, etc., for human motivation and behavior, systematically relegates transcendent experiences, of whatever religious persuasion, to the level of unresolved neurotic conflicts, such as an infantile desire for symbiosis with the mother, the need for a father figure, magical thinking invading ego functions, etc. Belief in God is thus, equated with neurosis. The good news is that this type of psychopathology is considered to have a good prognosis, given proper treatment. Even though the field of psychoanalysis has traditionally challenged the simplistic explanations of conscious experience inherent in the medical model of psychiatry, it has itself fallen victim to it with respect to the issue of God and religion. Is it possible that this attitude constitutes a collective defense mechanism in the psychoanalytic community stemming from its need to be more acceptable in the eyes of establishment psychiatry?

5.3. "PARANORMAL" PHENOMENA

Scientific bias obscuring objectivity also reigns in the field of paranormal phenomena, but assumes a different form than that for 'near death' experiences. When it comes to investigating near death experiences, materialist assumptions about the nature of reality confine the nature of scientific investigation to materialist mechanisms of brain physiology that might explain the phenomenon. When it comes to paranormal phenomena, this same principle eliminates the discussion altogether by denying that such events have ever occurred. The denial is interesting because there have been many experiments of "paranormal" phenomena during the past few decades (e.g., at the Princeton Engineering Anomalies Research Lab) which scientists simply don't bother to read. Since there is no plausible mechanism within a materialist frame of reference to explain them, paranormal phenomena can't possibly exist and there is no need to examine data implying the contrary because it can't be valid. This is the same reasoning that the learned men of Galileo's day used when they refused to look in the telescope.

This attitude is nowhere more evident than in the number of scientists who are willing to volunteer as "expert" commentators on television programs about paranormal phenomena, astonishingly undeterred and unembarrassed by their complete lack of knowledge concerning the existing experimental data. These "experts" smile condescendingly as they explain that the phenomena under discussion can be explained by chance occurrence, brain abnormality, etc., depending on the topic at hand. Since the belief that causality can only be found in matter reigns supreme, there doesn't seem to be any requirement that these "experts" support their claims with actual data. They need only introduce the possibility that the same outcome might have been achieved through some other means, to be convince their naïve audience that it is all 'hocus pocus'.

The refusal on the part of most mainstream scientists to even consider paranormal phenomena as a legitimate area for scientific investigation is unfortunate because it leaves a wide open market for charlatans, who have no qualms about making a living off of the needs of gullible people. However, it is also unfortunate for another reason. Refusal to acknowledge this area as a legitimate focus of scientific investigation is limiting our knowledge of human consciousness and human potential. The resulting dearth of active research fuels the continued refusal to examine the existing data and the cycle of ignorance continues unabated. The aim of this section of the book is to discuss some of the most recent data concerning paranormal phenomena, along with the physical principles that are postulated as mechanisms. The purpose is to demonstrate that these data should be taken seriously enough by the scientific community to stimulate further investigation.

For more than two decades, the Princeton Engineering Anomalies Research Lab (PEAR), headed by Professor of Aerospace Sciences and Dean Emeritus of the School of Engineering and Applied Science at Princeton University, Robert Jahn, has conducted experiments to investigate whether humans can influence random event generators or employ precognitive remote perception [5.53]. Random event generators (REG) generate an electronic source of "white noise." The circuitry then samples this noise signal at regular intervals, translating the signal into a "plus" if that particular signal is higher than the mean value or a "minus" if it is lower than the mean value. The result is a random string of pluses and minuses displayed on a visual display board. The individuals participating in the experiments are instructed to try to mentally influence the sequence of pluses and minuses by any strategy they think will work. In some series they are instructed to try to increase the number of pluses, in some, to increase the number of minuses. The machine displays continual feedback. This experiment resembles the flipping of a coin. If a coin is flipped hundreds of times, by chance it should come out with a fairly equal number of heads and tails. The reason for using a random events generator instead of a coin is that the individual never touches the generator and therefore, cannot influence it physically. If the individual is flipping coins and trying to mentally influence them, s/he may inadvertently throw them differently. The REG eliminates that possibility. A similar series of experiments was performed with another machine called a random me-chanical cascade (RMC). This is a tall machine with a bunch of little balls that are also propelled outward in a random way and fall downward into dif-ferent holes. The accumulated data from thousands of experiments with the-se two machines concludes that human operators can influence their output in statistically replicable and individually characteristic ways [5.53].

The other type of experiment performed at PEAR, is called precognitive remote perception. In these experiments, one individual is assigned to "re-ceive" and another to "send" the image of a defined location, which is not known to the sender until s/he arrives there. The 'receiver' stays in the lab and the sender is told to drive to a specific location. At that location he or she then receives instructions to go to further destinations. At a specific point in time, the sender is asked by the experimenter to concentrate on the scene in front of him/her and "send" that image to the receiver. The individual who is to "receive" is asked to describe *ahead of time* (i.e., before the sender is instructed to stop and observe), what that the sender will be sending within the next couple of hours. His or her verbal description is recorded, and an-swers to an additional 30 detailed questions about the content of the scene, requiring "yes" or "no" answers, are also quantified. Examples of the ques-tions are: "is the lighting dim or dark?" "Are animals visible at the location," and etc. The same 30 questions detailing as many environmental aspects as possible are asked of each individual and the number of right and wrong an-swers are recorded. In their book, Jahn and Dunne show photos of the loca-tions. The taped descriptions are sometimes uncannily accurate. The con-clusion drawn from these experiments is that "substantive information about geographical targets that is inaccessible by any known sensory channel, has been acquired by remote percipients, with a degree of fidelity that appears to be statistically insensitive to the intervening space or time" [5.54]. One inter-esting aspect of these experiments is that they have also been performed be-tween countries. The sender in these cases is someone who goes to another country on business or vacation and agrees to "send" from that country at a predetermined time during the trip. What is predetermined is the time but not the location of the image that will be sent. Some of these trips involved countries to which the sender had never traveled, and for which s/he could therefore have had no clear mental pictures related to specific locations before departure. Nevertheless, the receiver has many times accurately described the scene, even before the sender picked it out. Impossible? Since the phenome-non has been repeatedly verified, it is not impossible. The question must change to, "what is the explanation?"

According to Einstein's theory of relativity, which has been validated by every experiment that has tested it, the speed of light is constant. However, space and time are not. Time slows and space diminishes as speed increases. Upon reaching the speed of light, space, time and matter disappear altogether. This means that there is no distance between the point of emission of a photon and its point of absorption, and that there is no time but the present. Let us recall what quantum theoretical experiments have repeatedly demonstrated about non-locality, namely that two particles created from the same photon but separated by space can, under certain circumstances, be continually aware of measurements made on each other despite not having any possibilities for sending or receiving "messages." Both the concepts of relativity and non-locality have enormous implications for precognition, telepathy and intercessory prayer. Since quantum theoretical experiments have clearly demonstrated that matter is non-local, why would we assume that mind does not also display the same characteristic? This was also the conclusion drawn by surgeon, Wilder Penfield, who concluded during 50 years of neurosurgery in which he stimulated every conceivable area of the cortex, that consciousness was not localized within the brain. Could it be that precognition as we define it is not precognition at all, but a form of consciousness (non-matter) that exists within a constant present? Telepathy and intercessory prayer may not involve "sending" messages but thinking in a timeless space.

The authors of a book on the Princeton experiments, Jahn and Dunne, make another very important point, namely, that consciousness, defined by all of our thoughts, feelings, emotions, perceptions, intuitions, etc., has also conceived the ordering principles that we use to conceptualize scientific theories:

> the consciousness that has conceived both particles and waves, and found it necessary to alternate them on some complementary fashion for the representation of many physical phenomena, may find a similar complementarity necessary and useful in representing itself. . . .the commonly prevailing conceptualization of consciousness has been basically 'particulate' in nature. That is, an individual consciousness has usually been presumed to be well localized in physical space and time, namely in its host physiological corpus, and to interact only with a few aspects of its environment and a few other similarly localized consciousness at any given point in its experience. . . .if consciousness were to allow itself the same wave/particle duality that it has already conceded to numerous physical processes, these situations would become more tractable. In particular, a wave-like consciousness could employ a host of interference, diffraction, penetration, and remote influence effects to achieve normally most of the anomalies of the particulate paradigm.[5.55]

Unlike particles, wave forms are not altered by collision. They can pass right through each other, forming interference patterns that leave no permanent distortion. They can also diffract around corners. When they meet a discontinuity in the medium, some of them are reflected and others continue to be transmitted, unlike particles which are completely deflected. According to Jahn and Dunn's theory, waves of consciousness comply with physical properties and laws. Thus, velocity, which is the rate of change of position per unit time, finds its analogue in consciousness as the rate of acquisition of information. A high rate of acquisition would be characterized by a short wavelength, a slow rate of acquisition, by a longer wavelength. If physical kinetic energy is defined as the energy possessed by an object based on its motion, conscious kinetic energy is the capacity for producing changes in consciousness itself or the surrounding environment. According to this concept, higher energy has greater capacity for overcoming environmental barriers. The authors define in technical terms, different vectors of the coordinate system: the range of penetration of the environment, the emotional component, and the orientation or perspective of consciousness. Their observation from their own and others' work is that emotional or physiological bonds between sender and receiver increase the accuracy of precognitive remote perception. Although they have never specifically tested this, it is an interesting observation that relates back to the discussion of intercessory prayer and the implications of an emotional connection to the patient. Jahn and Dunn also refer to the theory of relativity which shows that there is mathematically little difference between spatial and temporal behavior since they both change relative to changes in velocity, diminishing as the speed of light is approached. Thus, any mechanism that can explain getting information from a remote distance should be applicable to information that is remote in time.

At this point, I would like to return to the concept of non-locality described in the section on "emergent properties" because I think it has relevance for how we think about consciousness. In that chapter, I suggested that non-locality could be conceptualized by comparing the photons which are separated by miles but act as if they are in constant *simultaneous* communication with each other, to two ocean waves propagated in opposite directions but connected by the same ocean. This analogy is somewhat similar to what Carl Jung described as the Collective Unconscious. Jung was part of the early psychoanalytic movement and at one time a colleague of Freud, who later was expelled ('excommunicated') from Freud's circle because he developed very different ideas about human consciousness. Jung came to the conclusion through his psychoanalytic work, that individual human consciousness is connected to, or part of, a greater Collective Unconscious, i.e., that the individual is part of a greater whole that includes all other human beings. He based this theory on his repeated observations that individuals with no connection to each other exhibited evidence of what he referred to as universal "archetypes." These archetypes are similar thought forms or symbols that appear in the thoughts and fantasies of all individuals, regardless of culture. In Jung's opinion, these archetypes manifest in individuals but exist prior to individual consciousness. Jung's definition of the mind involved an "open" system, with exchanges occurring between the individual and the collective unconscious. Freud's concept of the mind, on the other hand, is a closed system. It is based on physical drives, such as sexuality, that are hardwired into the system. According to Freud, psychic energy stemming from these drives resides in the id and is transferred to the ego and the superego, which help the id to satisfy or, when necessary, to sublimate its drives. No energy can escape Freud's closed system of psychic energy, but it can be shunted between the three levels. It can be transferred back and forth between id, ego, and superego, but is confined within the individual (by implication in the brain) and does not dissipate. Thus, it involves a "zero sum" game.

Jung's theory of the collective unconscious did not sit well with Freud, and Jung has since also fallen out of favor with mainstream psychoanalytic theory, except for a special branch, who call themselves "Jungians." Although Freudian psychoanalytic theory has consistently maintained the importance of unconscious process as a source of motivation and behavior, its emphasis on physical drives as the source of psychic energy is consistent with the materialist point of view. Freud did not consider the id, ego and superego to be just theoretical constructs but envisioned them as actually existing in the brain. If they were not contained in the brain, then the system of energy could not have been closed. However, unconscious process, which refers to feelings and beliefs that exist outside our conscious awareness, has no neuroanatomical substrate. The id, ego and superego have no known physical correlates in the brain. So it is difficult to conceive of energy moving within a closed system between areas of the brain that either don't exist or haven't yet been identified. This puts contemporary psychoanalytic theory in a rather untenable position, or a sort of scientific limbo. On the one hand, the concepts of id, ego and superego and the energy associated with them have been clinically very useful in explaining human behavior and resolving neurotic conflicts. On the other hand they don't exist on a material level, even though materialism is what causes psychoanalysts to reject transcendent (nonmaterial) concepts as invalid.

To summarize, modern psychiatry is based on a model of matter ; sality for psychological experience. Its belief in the ultimate primacy c.__ ter but simultaneous lack of knowledge concerning the physical attributes of matter, demonstrate ignorance of the nature of reality as expressed by the principles of modern physics. Psychoanalytic theory, with its strong emphasis on early experience, rather than chemical receptors as the cause of psychological experience, diverges in many ways from the medical model of psychiatry while imitating it in others. Jung's concept of the connectedness of individual consciousness to a greater "whole" (the Collective Unconscious), which is ridiculed by both the medical model of psychiatry and by mainstream psychoanalytic theory, is more consistent with the non-locality principle of quantum mechanics and with its concept of multiple simultaneous probabilities, then either of the other two viewpoints. Together, data on the nature of quantum reality, the existence of non-material paranormal phenomena, and the abundant psychophysiological studies detailing the ways that psychological experience influences brain chemistry, indicate the gross inadequacy of a unidirectional materialist approach to explaining mind/brain interactions and human consciousness. In light of these facts, the relative complacency of the scientific community towards the issue of consciousness is astounding. Consciousness is one of the least explored scientific frontiers and one of its greatest and most important enigmas.

REFERENCES

5.1)   Penfield W. *The Mystery of the Mind.* Princeton University Press, Princeton, New Jersey, 1975, pp 76-80..

5.2)   Penfield W. The electrode, the brain and the mind. *A Neurol* 1972;201: 297-309.

5.3)   Hofmann HA, Benson ME, Fernald RD. Social status regulates growth rate: Consequences for life-history strategies. *Proc Natl Acad Sci U S A* 1999;96:14171-14176.

5.4)   Fox HE, White SA, Kao M, Fernald RD. Stress and dominance in a social fish. *J Neurosci* 1997;17:6463-6469.

5.5)   Francis RC, Kiran S, Fernald RD. Social regulation of the brain-pituitary-gonadal axis. *Proceedings of the Proc Natl Acad Sci U S A* 1993;90:7794-7798

5.6)   Winberg S, Winberg Y, and Fernald RD. Effect of social rank on brain monoaminergic activity in a cichlid fish. *Brain Behav Evol* 1997;49:230-236.

5.7) White SA, Fernald RD. Changing through doing; behavioral influences on the brain. *Recent Prog Horm Res* 1997;52:455-473

5.8) Shively CA, Laber-Laird K, Anton RF. Behavior and physiology of social stress and depression in female cynomolgus monkeys. *Biol Psychiatry* 1997;41:871-882.

5.9) Williams JK, Shively CA, Clarkson TB. Determinants of coronary artery reactivity in premenopausal female cynomolgus monkeys with diet-induced atherosclerosis. *Circulation* 1994;90:983-987.

5.10) Kaplan JR, Pettersson K, Manuck SB, Olsson G. Role of sympathoadrenal medullary activation in the initiation and progression of atherosclerosis. *Circulation* 1991;84(Suppl VI):VI-23-VI-32.

5.11) Kaplan JR, Adams MR, Clarkson TB, Manuck SB, Shively CA, Williams JK. Psychosocial factors, sex differences, and atherosclerosis; lessons from animal models. *Psychosom Med* 1996;58:598-611.

5.12) Yokel RA. The toxicology of aluminum in the brain: a review. *Neurotoxicology* 2000;21:813-828.

5.13) Kalmijn S, Mehta KM, Pols HA, Hofman A, Drexhage HA, Breteler MM. Sublclinical hyperthyroidism and the risk of dementia. The Rotterdam study. *Clin Endocrinol* 2000;53:733-737.

5.14) El-Khodor BF, Boksa P. Transient birth hypoxia increases behavioral responses to repeated stress in the adult rat. *Behav Brain Res* 2000;107:171-175.

5.15) Mendez MF. The neuropsychiatry of multiple sclerosis. *Int J Psychiatry Med* 1995;25:123-130.

5.16) Weissman MM, Bland RC, Canino GJ, Faravelli C, Greenwald S, Hwu HgG, Joyce Pr, Karam EG, Lee Ck, Lellouch J, Lepine JP, Newman SC, Rubio-Stipec M, Wells JE, Wickramaratne PJ, Wittchen H, Yeh EK. Cross-national epidemiology of major depression and bipolar disorder. *JAMA* 1996;276:293-9.

5.17) Franke L, Schewe HJ, Muller B, Campman V, Kitzrow W, Uebelhack R, Berghofer A, Muller-Oerlinghausen B. Serotonergic platelet variables in unmedicated patients suffering from major depression and healthy subjects: relationship between 5HT content and 5HT uptake. *Life Sci* 2000;67:301-305.

5.18) Racagni G, Brunello N. Physiology to functionality: the brain and neurotransmitter activity. *Int Clin Psychopharmacol* 1999(Suppl 1):S3-S7.

5.19) Meyers S. Use of neurotransmitter precursors for treatment of depression. *Alternative Medicine Review* 2000;5:64-71.

5.20) Shapira b, Newman ME, Gelfin Y, Lerer B. Blunted temperature and cortisol responses to ipsapirone in major depression: lack of enhancement by electroconvulsive therapy. *Psychoneuroendocrinology* 2000; 25:421-438.

5.21) Yatham LN, Liddle PF, Shiah IS, Scarrow G, Lam RW, Adam MJ, Zis AP, Ruth TJ. Brain serotonin2 receptors in major depression: a positron emission tomography study. *Arch Gen Psychiatry* 2000;57:850-858.

5.22) Higley JD, Suomi SJ, Linnoila M. CSF monamine metabiolite concentrations vary according to age, rearing, and sex, and are influenced by the stressor of social separation in rhesus monkeys. *Psychopharmacology* 1991;103:551-556.

5.23) Shively CA, Laber-Laird K, Anton RF. Behavior and physiology of social stress and depression in female cynomolgus monkeys. *Biol Psychiatry* 1997;41:871-882.

5.24) Lanfumey L, Pardon MC, Laaris N, Joubert C, Hanoun N, Hamon M, Cohen-Salmon C. 5-HT1A autoreceptor desensitization by chronic ultra mild stress in mice. *Neuroreport* 1999;10:3369-3374.

5.25) Emerson AJ, KappenmanDP, Ronan PJ, Renner KJ, Summers CH. Stress induces rapid changes in serotonergic activity: restraint and exertion. *Behav Brain Res* 2000;111:83-92.

5.26) Bennett AJ, Lesch KP, Heils A, Long JC, Lorenz JG, Shoaf SE, Champous M, Suoi SJ, Linnoila MV, Higley JD. Early experience and serotonin transporter gene variation interact to influence primate CNS function. *Mol Psychiatry* 2002;7:118-122.

5.27) Meaney MJ, Szyf M. Maternal care as a model for experience-dependent chromatin plasticity. *Trends in Neurosci* 2005;28:456-463.

5.28) Fournier JC, DeRubeis RJ, Hollon SD, Dimidjian S, Amsterdam JD, Shelton RC, Fawcett J. Antidepressant drug effects and depression severity. *JAMA* 2010;303:47-53.

5.29) Caspi A, Sugden K, Moffitt TE, Taylor A, Craig IW, Harrington H, McClay J, Mill J, Martin J, Braithwaite A, Poulton R. Influence of life stress on depression: moderation by a polymorphism in the 5-HTT gene. *Science* 2003;301:386-389.

5.30) Moncrieff J, Wessely S, Hardy R. Meta-analysis of trials comparing antidepressants with active placebos. *Br J Psychiatry* 1998; 172:541-542.

5.31) Antonuccio DO, Danton WG, DeNelsky GY, Greenberg RP, Gordon JS. Raising questions about antidepressants. *Psychother Psychosom* 1999;68:3-14.

5.32) Suomi SJ, Harlow HF, Domek CJ. Effects of repetitive infant-infant spearation of young monkeys. *J Abnorm Psychol* 1970;76:161-172.

5.33) Suomi SJ, Delizio R, Harlow HF. Social rehabilitation of separation-induced depressive disorders in monkeys. *Am J Psychiatry* 1976;133:1279-1285.

5.34) Suomi SJ. Early determinants of behaviour: evidence from primate studies. *Br Med Bull* 1997;53:170-184.

5.35) Agid O, Shapira B, Zislin J, Ritsner M, Hanin B, Murad H, Troudart T, Bloch M, Heresco-Levy U, and Lerer B. Environment and vulnerability to major psychiatric illness: a case control study of early parental loss in major depression, bipolar disorder and schizophrenia. *Mol Psychiatry* 1999;4:163-172.

5.36) Suomi SJ. Anxiety-like disorders in young primates. In; gittelmann R (ed) *Anxiety Disorders of Childhood*. New York: Guilford, 1986;1-23.

5.37) Grinshpoon A, Mester R, Spivak B, Berg Y, Bleich A, Weizman A. Delayed amnesia an disorientation after electroconvulsive treatment. *J Psychiatry Neurosci* 1992;17:191-193.

5.38) Calev A, Nigal D, Shapira B, Tubi N, Chazam S. Ben-Yehuda Y, Kugelmass S, Lerer B. Early and long-term effects of electroconvulsive therapy and depression on memory and other cognitive functions. *J Nerv Ment Dis* 1991;179:526-533.

5.39) Chamberlin E, Tsai GE. A glutamatergic model of ECT-induced memory dysfunction. *Harv Rev Psychiatry* 1998;5:307-317.

5.40) Lewis P, Kopelman MD. Forgetting rates in neuropsychiatric disorders. *J Neurol Neurosurg Psychiatry* 1998;65:890-898.

5.41) Lisanby SH, Maddox JH, Prudic J, Devanand DP, Sackeim HA. The effects of electroconvulsive therapy on memory of autobiographical and public events. *Arch Gen Psychiatry* 2000;57:581-590.

5.42) Donahue AB. Electroconvulsive therapy and memory loss: a personal journey. *J ECT* 2000;16:87-96.

5.43) McCall WV, Reboussin DM, Weiner RD, Sackeim HA. Titrated moderately suprathreshold vs fixed high-dose right unilateral electroconvulsive therapy: acute antidepressant and cognitive effects. *Arch Gen Psychiatry* 2000;57:438-444.

5.44) Shapira B, Tubi N, Lerer B. Balancing speed of response to ECT in major depression and adverse cognitive effects: role of treatment schedule. *J ECT* 200;16:97-109.

5.45) Devaney RL. Chaotic explosions in simple dynamical systems. In S. Krasner (ed). *The Ubiquity of Chaos* American Association for the Advancement of Science, Washington, D.C., 1990.

5.46) Sabom M. *Light and Death.* Zondervan Publishing House, Grand Rapids, MI, 1998.

5.47) Ring L. A note on anesthetically-induced frightening "near-death experiences." *J Near-Death Studies* 1996;15:17-23.

5.48) The ketamine model of the near-death experience: A central role for the N-methyl-d-aspartate receptor. *J Near-Death Studies.* 1997;16:5-26.

5.49) Serdahely WJ. Similarities between near-death experiences and multiple personality disorder. *J Near-Death Studies* 1992;11:19-38.

5.50) Irwin HJ. The near-death experience as a dissociative phenomenon: An empirical assessment. *J Near-Death Studies.* 1993;12:95-103.

5.51) MacHovec F. Near-death experiences:; Psychotherapeutic aspects. *Psychotherapy Priv Pract.* 1994;13:99-105.

5.52) Greyson B. The near-death experience as a focus of clinical attention. *J Nerv Ment Dis.* 1997;185:327-334.

5.53) Jahn RG, Dunne BJ. *Margins of Reality.* 1987, Harcourt Brace & Company, New York, N.Y. pp. 92-102 and pp. 149-191.

5.54) Ibid. p. 195.

5.55) Ibid. pp. 211

CHAPTER 6

♦

WHAT IS LIFE?

Psychophysiology investigates the way that psychological experience is translated into physiological response patterns and psychobiology does the opposite, i.e., it investigates the way in which biological processes influence the psyche. Both describe the neuroanatomical, neurophysiological and neuroendocrine responses associated with specific psychological experiences (e.g., stress, well-being, fear, depression, etc.). Thus, although approaching the subject matter from different perspectives, they bridge two dimensions: the psychological and the physiological. My hypothesis, based on the evidence presented, is that there are two additional dimensions needed to complete our understanding of mind / body interactions: an energy dimension, or biophysical aspect; and a spiritual dimension, which is inextricably intertwined with the other three, and is the essence of who we are.

Up to this point, I have focused primarily on the biophysical dimension, i.e., the implications of quantum mechanics and endogenous force fields for living systems. This has been necessary because even though the data supporting this hypothesis are not new, they are relatively unknown to the biomedical community and have never been compiled in one place to illustrate their relevance to mainstream clinical medicine. I believe that adding biophysics to the study of biological mechanisms would eliminate many of the discrepancies between modern physics and current views in the biological and medical sciences. I have hypothesized a mechanism that might serve as a physics / biology interface, namely energy meridians in the form of dipole chains. In discussing this data, I described my willingness, indeed my earlier desire, to keep the discussion on a biophysical level. But I have come to believe that these principles are insufficient for a comprehensive understanding of the nature of reality. The mechanical transfer of energy from one person to another is indeed possible, but it explains only a small part of the variance in the equation. The state of consciousness of a healer is part of the healing interaction. The surface potential measured by the EEG is an indication that thoughts are associated with energy patterns. If this is the case, they are also associated with a force field that changes (if neuroendocrine secretions are a reliable indication) with emotion. It follows from this that not only the subjective state of the healer but also that of the recipient may influence the outcome. This is undoubtedly part of what has long been labeled the placebo effect in traditional medicine. However, it indicates that rather than controlling for, or eliminating subjective states through rigorous experimental method, we should be including them in the experimental design in order to rigorously investigate their effects.

If this line of reasoning is correct, it also follows that telepathy experiments are influenced by the mental states of the sender and receiver. For instance, repeating the same experiment over and over during a single session, in order to test "reliability" can become so boring for the sender that s/he has increasing difficulty focusing and sustaining concentration. Bored participants lacking in concentration would be expected to do poorly compared to participants whose concentration is focused on the experiment. Understanding this, allows it to be taken into consideration when designing the study. It also has implications for how one would design an investigation of intercessory prayer. A person praying for a bunch of names on a list, knowing nothing about these people as individuals is very likely to have a completely different state of mind and energetic force field (pattern of surface potentials emanating from the brain) than someone praying for an individual whose status and progress are known. The latter has become a person for whom the prayer feels support and empathy, which are likely to generate a much stronger emotional connection (and thus force field) than the individual praying for an anonymous list of people whose condition is unknown.

These reflections concerning the influence of subjective states on experimental results lead to the question of what consciousness actually is. In the medical model of psychiatry, it is defined as the product of the sum total of the organic brain, i.e., of particulate matter such as cells, receptors, neurochemicals, genes, etc. In the chapter on consciousness, I gave some examples for which this model seems inadequate, including telepathy, near death experiences, and intercessory prayer. In that discussion, I attempted to broaden the concept of consciousness by describing the applicability of quantum theory and wave / particle duality. In this context, I also described the principle of non-locality, acknowledging that although its tenets have been unequivocally verified, even physicists still have difficulty comprehending them. I made an attempt at clarification by using the analogy of separate waves being part of the same ocean, and drawing parallels to Jung's concept of the Collective Unconscious.

To develop this idea still further, it would be pertinent to mention more about the theories associated with acupuncture. The Chinese version is based not on principles of physics, but on the concept of a living energy they call, "chi (Qi)." According to this theory, the meridians are composed not of dipoles, but of living energy. When the chi flows naturally and without obstruction, there is balance, both physically and mentally. Blockages in the chi meridians can cause illness. Chi is not only an integral part of biological systems, it is everywhere in the universe. Feng Shui design of buildings and houses is said to improve the inhabitants' lives and fortunes through the principle of improving the 'flow of chi'. The concept of chi is not so far from the Christian belief in "spirit", although chi is not seen as representing a Divine Being. Chi simply is.

The tremendous, almost instantaneous surge of energy I felt from my hands when I entered a state of deep love while healing my daughter is more consistent with a living energy than with abstract dipole chains. This does not mean that the mechanism does not involve dipoles. What it means is that there seems to be something more. Clearly, these reflections are speculative, but I believe that science has arrived at a point where the data have outgrown the existing theoretical framework on which we have based our premises. If we are to make progress, we must begin to ask new questions.

Until recently, it was possible and even desirable to perpetuate a "separation of church and state" with respect to science, by keeping science isolated from philosophic domains. After all, we certainly don't want a repeat of what happened in Galileo's time, where the church dictated what one was permitted to investigate. However, even scientists who believe in God see no contradiction in leaving God outside the door when they walk into their laboratories. They are not in the least bothered by the fact that the mechanism of creating life from "non-life" (inert matter) has never been satisfactorily resolved. Darwin's theory of common descent did not explain this and science is still struggling with the issue.

The problems involved in this type of research can be exemplified by an interesting experiment which demonstrated that it is possible to create an RNA molecule in the laboratory under prebiotically plausible conditions [6.1], which means conditions that are thought to have existed on the earth before life. RNA is the molecule that translates DNA in order to create proteins which are the building blocks of everything in the body. The catalyst for these experiments was irradiation with ultraviolet light. What is interesting about this experiment is that it could theoretically have happened under the conditions that existed on earth at the time. But going from an RNA molecule to a living organism is a huge transition. We replicate DNA in laboratories all the time to make probes for microarray analyses (under conditions that are not prebiotic). In this process we cause the DNA to replicate itself in something called a polymerase chain reaction, by adding a chemical catalyst. However, these DNA replications that we create are not living organisms. They are templates or 'probes' that we put into dishes to help us identify the mRNA in our biological samples. They do not become anything else and none of us would define these replicating sequences as 'life'.

A more recent experiment focuses even more directly on the question of the essence of life. Scientists succeeded in assembling an entire synthetic genome and transplanting it into a recipient cell to reprogram the cell into a different type of cell [6.2]. This was an amazing technological feat and is being reported as an 'artificial life' breakthrough [6.3]. What the experiment did was to take the chromosomes (DNA) out of one type of bacterial cell and replace it with synthetically constructed DNA equivalent to the chromosomes of another bacterial cell. The cell then began to act like the bacteria consistent with the inserted DNA. So the question is what is life? Did this experiment create artificial life? The first author of the paper likened the process to making new software for the cell. He says, "as soon as this new software goes into the cell, the cell reads and converts [it] into the species specified in that genetic code". But what part of the cell is doing the 'reading' here? Clearly the cell did not die when the DNA was removed. So if DNA is the software programming 'life', we are still left with the question, what is life?

Let us for a moment do a thought experiment about the intersection of 'life' and the physical body. In this thought experiment, a man is stabbed in a dark alley at 2:00 a.m. The knife makes a clean cut in the main blood vessel that pumps blood from the heart. He is left face down in the alley and bleeds to death. Six hours after his death, his body is discovered. Medical science now has techniques to repair this type of injury. He can be taken to a hospital, have the artery surgically repaired, lost blood replaced with his exact blood type, and the cavity cleaned of any coagulated blood. The man would then have all the DNA, tissue and anatomical structures he had before he died, but he would still be dead. "Life" has left his body and will not return. One might respond that this is because his brain cells did not get oxygen and "died". Well, what does that mean exactly? What is a 'dead' brain cell? We know that people who are "brain dead" can have bodies that continue to survive for many years. We can even put our theoretical man on an artificial pump after surgery so that blood and oxygen are circulated through his brain and body. But he will not come back to life. What then is life? Is this not a highly relevant question for biomedical scientists to be asking? If we are to treat and prevent malfunction in living systems, do we not need to understand the essence of what constitutes 'being alive'? As the data from earlier chapters demonstrate, reality cannot be reduced to bits of matter and life cannot be explained by the sequence of DNA base pairs. The dead man has DNA but not life.

At a subatomic level, the difference between matter and energy disappears. If we continue along our current path of allowing materialism to serve as the fundament for biomedical research we will never find the answer to what life is because we will be looking in the wrong place. Materialism was a philosophical stance that was co-opted by the scientific community along with the integration of Newtonian mechanics. However, we are now in the 21st century and this paradigm has long since become outdated. We must either begin to integrate modern physics into the fundamental tenets of our hypotheses or risk becoming theoretically irrelevant. The fact that we can measure the part of the brain that is activated during certain types of cognitive processes (e.g. with blood flow, fMRI, evoked potentials, etc.), describe the hormonal secretions and neuroanatomical substrates associated with emotion, and know how impulses are transmitted between neurons has not brought us any closer to understanding what thought is. This should be a huge red flag that our methodology is somehow incomplete, but it seems to have escaped notice altogether.

If Jung's theory of the Collective Unconscious is accurate, then our methodology may be flawed. The waves cannot accurately measure the ocean. It is not possible to stand "outside" of what we are observing and be an objective observer because we are part of it. This is a fundamental principle of quantum mechanics, that the very act of measurement influences the results. This insight needs to be incorporated into the way we design experiments and interpret results in biomedical research.

Even the broader concept of a physics / biology interface does not seem to be complete because it too, excludes relevant data. One of the excluded data points is the strong evidence of a life after death, as exhibited in many "near death" experiences. Despite attempts by the medical and psychiatric communities to explain these experiences as epiphenomena caused by drugs or oxygen deprivation, this would not enable people to observe what is going on around them while they are unconscious. Nor do hallucinogenic drug experiences tend to permanently alter people's lives by giving them a deeper sense of purpose and commitment to living in a more meaningful or spiritual way (a common consequence of the ND experience). If they did, drug addicts would be the pillars of our communities, not the primary inhabitants of our prison populations. The current medical definitions of human consciousness and even the functioning of the human body contain only partial truths. If one objectively examines the ND data, one cannot exclude the possibility that these experiences may not be epiphenomena but an actual indicator that there is a life after death. If there is a life after death, then we must have a soul and if we have a soul, then we are spiritual beings experiencing a physical existence. This is so contrary to the current scientific definition of what constitutes the essence of being human that it will probably be met with a great deal of scorn and derision. However, any skepticism it may generate is useful because skepticism leads to more questions as well as to experiments designed to prove that the theory is wrong.

Our current beliefs about the essence of what it means to be human are precisely that – beliefs. As scientists we must be more reliant on data. The purpose of this book is to get the reader to examine some of the data that are currently being ignored and to stimulate a search for answers to the questions they evoke. The implications for broadening the current scope of research defining human consciousness, human potential, and physical and mental well-being could be profound.

Although phenomena such as pre-cognition could be attributable to the non-locality principle of quantum physics, that principle can only partially account for the effect of intercessory prayer (if the effect exists). If one postulates that thoughts have energy, then it might be possible to leave spirituality and even non-locality out of the equation by saying that intense thoughts of healing are somehow "transmitted" to the patient and that they can influence the physiology of the patient's body. If this were the case, it would certainly make sense to harness such tremendous "thought power" to heal. However, if we are all part of a larger, Collective Unconscious, then maybe non-locality (and healing) works like the waves talking to the ocean -- through an already existing spiritual connection, so that nothing is being transmitted at all. In this model, the healer and the person being healed would be parts of the same consciousness. Regardless of which stance you take, the implications open the door to profound new dimensions and a whole new concept of healing.

Belief systems help us feel secure because they help us to make sense of the world around us and to define our place in it. They can also impede us if they are not accurate. If materialism were an accurate way of describing the evolution and function of living systems in the universe, then God would, by definition, be unnecessary. The fact that we now know that materialism is not an accurate description of the nature of reality or of human mental and physiological functioning does not mean that God *is* necessary. However, it does destroy one of the basic scientific tenets that supposedly proved the non-existence of a Divine Being. From the standpoint of science, the question of God's existence is still an unresolved issue.

Does the existence of this question mean that it is a relevant focus of scientific inquiry? What could be more relevant? If God exists, then it would open the possibility that our true nature is that of spiritual beings temporarily manifesting or inhabiting physical bodies. The implications would be enormous. This would imply the possibility of a fourth, spiritual, dimension to human existence, in addition to the material, energetic, and psychological / emotional dimensions already described. It is a question that goes to the essence of what defines being human.

REFERENCES

6.1) Powner MW, Gerland B, Sutherland JD. Synthesis of activated pyrimidineribonucleotides in prebiotically plausible conditions. *Nature* 2009; 459: 239-242.

6.2) Gibson DG, Glass JI, Lartigue C, Noskov VN, Chuang R-Y, Algire MA, et al. Creation of a bacterial cell controlled by a chemically synthesized genome. *Science* 2010;329:38-9.

6.3) Gill V. 'Artificial life' breakthrough announced by scientists. BBC News; Available at: http://www.bbc.co.uk/news/10132762. Accessed May 20, 2010.

CHAPTER 7

THE SCIENTIFIC PURVIEW:
THE DEFINITION OF LEGITIMATE DATA

"The idea that observation can be pure and unsullied (and therefore beyond dispute) – and that great scientists are, by implication, people who can free their minds from the constraints of surrounding culture and reach conclusions strictly by untrammeled experiment and observation, joined with clear and universal logical reasoning – has often harmed science by turning the empiricist method into a shibboleth."

—Stephen Jay Gould [7.1]

The term legitimacy as I will be using it does not refer to lack of data falsification, but to the process that determines which data qualify as relevant topics of scientific inquiry. What the doctrine of materialist reductionism has *de facto* accomplished through its domination of the life sciences, is to declare certain types of data (e.g., electromagnetic phenomena in biological systems, transcendent experience, spiritual healing) as unscientific and therefore outside the legitimate purview of science. Alternatively, it permits them only under special circumstances – as exemplified by forms of treatment that may only be tried as "alternatives when all else has failed." This bias has created a situation where it is considered perfectly legitimate to use toxic chemicals to treat malignancies that have not been demonstrated to respond well to chemotherapy. However, it is not legitimate or even legal to recommend a non-toxic treatment, such as acupuncture, except as an adjunct to chemotherapy, or as a last resort when a patient is terminal – after chemo and radiation have weakened or destroyed the immune system. Of course, if one waits with acupuncture until the immune system is completely dysfunctional, the DNA damaged, and the cancer metastasized out of control, the possibility for success is appreciably diminished.

This brings us to the whole concept of "alternative medicine," a term that I find nonsensical. Either there is a reason to believe a particular treatment is efficacious because existing data show an association with functional mechanisms and/or clinical outcomes, or there isn't. If it works, it should be part of mainstream medical interventions. If it doesn't, it shouldn't be used at all. Based on what modern physics tells us about the nature of reality, as well as the evidence relating electromagnetic fields to important mechanisms in chemistry, developmental biology and physiological functioning, the evidence for an important role of bioelectric phenomena in organic life forms is overwhelming. Despite this fact, mainstream medical scientists have shown little or no interest in conducting research on possible clinical applications utilizing these mechanisms.

However, electromagnetic phenomena are only the tip of the iceberg when it comes to questions of legitimacy in science. In a sense, they are actually the easiest, because they can be tested with standard, already existing methodology. The area of research that has been most sadly neglected, and whose methodologically is more difficult, concerns the area of subjective experience, e.g., feelings, cognition, mental imagery, perception, creativity, etc. The misconception that all of these phenomena have their causality in and are reducible to biochemical phenomena in the brain, has greatly impeded the progress of psychiatry, psychology and the study of human consciousness. The rigidity of the materialist way of thinking has caused science to become phase locked into a very narrow range of hypotheses.

Although the most blatant example of this is the area of transcendent experience, the problems with our approach to investigating it can probably be more easily illustrated by examining a feeling that is almost universal to human experience - love. There are many different kinds of love – romantic, platonic, maternal / paternal, and the feelings of universal love described by many mystics – but they have in common an intensity of feeling associated with the loved person's inner being, often in spite of any outward appearance. Love is also associated with the willingness to sacrifice one's own immediate needs, or even life, in favor of the person who is loved. There are scientists who would have us believe that this is just a result of biochemistry. Although there are indisputably areas of the brain and certain neurohormones that are associated with positive feelings, it is not possible to know what a person is feeling by measuring circulation in those specific areas of the brain or the associated hormone levels. People who are sexually aroused have elevated heart rate but so do people who have been running or people who are angry. These physiological signs are what we call "biomarkers," i.e., biological signs that may accompany these feelings just as the wheels of a car rotate when a car accelerates uphill. The two are correlated but not causally related. If love could be reduced to a hormone, I believe we would long ago have found a pill for it. However, the complexity of love is one of the deepest most intense and noble feelings that human beings experience and has always defied reductionist definitions. Although responses to certain gender and sexually based stimuli do seem to be programmed into all species of animals, sex is not equivalent to love. The love of a person for a spouse who is dying of cancer has long since lost any sexual content. The person's body had deteriorated to almost nothing and their appearance is diametrically opposed to any cultural perceptions of what is sexually attractive. However, the love is intense and the pain associated with the prospect of losing that person, excruciating. What exactly is love?

What about "maternal instinct?" This cannot simply be 'hardwired' into the brain because the ability to nurture a child varies so much within and between species. The ability to nurture seems to involve complex gene / environment interactions in both animals and humans. Non-human primates in captivity who have grown up isolated from maternal role models have not automatically known how to take care of their young. However, it has been demonstrated that neglectful behavior can be modified or avoided with the introduction of role models who demonstrate infant care. In humans, there are biochemical changes accompanying birth and lactation, which have the potential for creating positive reinforcement of proper mothering. However, these biochemical responses are not sufficient to create a strong mother / child bond or there wouldn't be mothers who fail to bond, abuse their children, etc. So even though love can be rationalized as a way for the species to survive, this is a *post hoc* attribution of causality and not proof that love can be reduced to something hardwired into our brains. Just as hormonal variation can result in mood swings, psychological mood changes can cause physiological changes [7.2-7.5], demonstrating again that causality is complex and multi-dimensional. Although love is universal to human experience, we have not succeeded in understanding its essence or its complexity.

If we cannot understand something as universal to human experience as love, how are we to comprehend experiences such as "enlightenment" or "near death" phenomena? Writing them off as chemically induced hallucinations is demonstrably inadequate and shows a paucity of intellectual depth. Rather than being scientific, that approach seems more consistent with a psychological defense mechanism bordering on denial - an impoverished way of coping with data that don't fit our hypotheses.

What then is the solution? One option is to at least include in the realm of possibility, that people relating transcendent experiences, such as "near death" encounters, may have actually experienced what they say they have experienced, namely another dimension beyond our earthly existence. If someone goes to a doctor and says, "I have been bitten by a snake," the doctor includes in the realm of possibilities, that the person may actually have been bitten by a snake. The patient may also be lying, imagining it, or even hallucinating. But he may actually have been bitten by a snake. Doesn't scientific objectivity require investigation of the possibility that the patient may be telling the truth? And aren't scientists supposed to be objective?

The point I am trying to make is that the problem is not that scientists have biases. Bias is an inherent characteristic of the human condition. The problem is rather, that scientists are unaware of their biases and therefore continue to perpetuate them. When the majority of scientists have the same bias, the risk is high that the progress of science will be severely impeded. Recall the learned men of Galileo's time who refused to look in the telescope. They were of the opinion that data from telescopes was not relevant. The same thing is happening today, except that the limiting doctrine is not coming from the Catholic Church. It is coming from science – the new religion of the 21st century. The dogma of this new religion is as rigid as that of the earlier church in dictating what is and is not acceptable in the scientific purview. Interestingly, a recent issue of a prestigious scientific journal carried an article with the title, "Plans for London 'Cathedral of Science' Unveiled" (7.6). We have finally come full circle. Scientists have, for so long, perpetrated the myth of their own objectivity that they have begun to believe it themselves. If those of us involved in research applied the same scrutiny to the nature of our scientific questions that we apply to statistical analyses, much scientific progress could be achieved.

This book is a challenge to the scientific community to begin to question the fundamental assumptions of materialism currently underlying scientific inquiry in the life sciences. Since the existence of a Divine Being cannot be eliminated with the extant data, the possibility that humans have a soul and the implications this would have for human existence should be considered a legitimate purview of scientific inquiry. The urgency of this issue can be found in its implications. If a God exists, then the principles of cosmology and human evolution may have been put in place by design. What does this mean for the evolution of human consciousness and the role of human beings in the context of the rest of creation? What is the role of science if not to investigate the questions that have the most fundamental importance for human existence?

REFERENCES

7.1) Gould J. *The Lying Stones of Marrakech.* 2000, Harmony Books, NY:31.

7.2) Kimbrell TA, George MS, Parekkh PI, Ketter TA, Podell DM, Daneilson AL, Repella JD, Benson BE, Willis MW, Herscovitch P, Post RM. Regional brain activity during transient self-induced anxiety and anger in healthy adults. *Biol Psychiatry* 1999;46:454-465.

7.3) Sargent CA, Flora SR, Williams SL. Vocal expression of anger and cardiovascular reactivity within dyadic interactions. *Psychol Rep* 1999;84(3 Pt 1):809-816.

7.4) Dougherty DD, Shin LM, Alpert NM, Pitman RK, Orr SP, Lasko M, Macklin ML, Fischman AJ, Rauch SL. Anger in healthy men: a PET study using script-driven imagery. *Biol Psychiatry* 1999; 46:466-472.

7.5) Burns JW, Evon D, Strain-Saloum C. Repressed anger and patterns of cardiovascular, self-report and behavioral responses: effects of harassment. *J Psychosom Res* 1999;47:569-581.

7.6) Travis J. Plans for London 'Cathedra of Science unveiled. *Science* 2009;326:1468-1469.

CHAPTER 8
◆
CONCLUSION

"In truth, science and religion present no contrasts but rather for every person of serious bent they need each other for mutual supplementation. It is surely no accident that precisely the greatest thinkers of all times were also deeply religious... Only out of the concurrence of the powers of the mind and of the will has there evolved philosophy's ripest, most precious fruit: ethics."

—Max Planck [8.1]

In the introduction, I stated that my belief in God did not come about through religion, but evolved as my deepening scientific knowledge made non-belief too difficult to sustain. The purpose of this book is to initiate a dialogue in the scientific community concerning the frame of reference that currently dominates and defines biomedical research. I stated that the existence or non-existence of God was an important scientific issue because scientifically (e.g., from a standpoint of existing data), it is unresolved and because our beliefs about the origin of the universe, life and the nature of reality influence the frame of reference we use to formulate scientific questions. The fact that a belief in materialism continues to dominate scientific inquiry in the fields of chemistry, medicine, psychiatry, psychology and biology, despite robust data demonstrating its inaccuracy, has led to major omissions with regard to the type of data included in research hypotheses. We have focused on material aspects of disease etiology and treatment and totally ignored the energy side of the equation. In so doing we have satisfied ourselves that the proverbial tusk of the elephant explains its essence and that the bases in the DNA double helix somehow suffice to explain physiological function and indeed, life itself. By ignoring the information available to us from modern physics, we have greatly limited our ability to understand and achieve success in major areas of disease pathology such as cancer treatment and mental illness.

What I hope to have accomplished with this book, is to show that the current scientific assumption of the non-existence or irrelevance of God, is really only a belief system masquerading as science. Despite the lack of supportive data, it influences the framework of scientific inquiry, including the formulation of scientific questions, the methodology, the variables we include in experiments, and the way we interpret data. For this reason, it needs to be closely examined. Materialism is based on theories that were once consistent with available scientific evidence, but became outmoded with the experimental data substantiating the tenets of quantum mechanics. The belief that matter is the sole cause of everything we see and know, and its accompanying belief, reductionism, which posits that we can achieve understanding of complex systems by studying their smallest constituent parts, are verifiably inaccurate. Yet, both of these beliefs continue to dominate research in the biological sciences. We cannot extrapolate the world of complex systems from small "building blocks" of matter because at a subatomic level such fundamental building blocks cease to exist and are replaced by a wave / particle duality and multiple potentiality. This means that energy characteristics are as important, if not more important, than particles of matter. It is not surprising that attempts to extrapolate the process of evolution based on a theory that somehow a living molecule appeared from inanimate matter, and went through random, surviving mutations, have proven grossly inadequate for explaining the depth of organic complexity we see around us and the essence of the human mind and consciousness. This is not to say that survival of random mutations has not been a part of the process, but that natural selection is an insufficient explanation for the complexity of evolution.

As scientists we have long perpetrated a myth to the general public that we are objective. This is decidedly not the case. We, like everyone else, have our biases and these biases have a penetrating influence on how we conduct research. It is time to begin an honest and open discussion of scientific beliefs, including the implications they have for our assumptions relating to the existence or non-existence of God. Since data are not consistent with a random, survival of the fittest explanation as the single driving force in evolution, we must be open to broader possibilities. The fact that current scientific data do not refute the existence of God does not constitute proof that God exists. However, it does leave the issue unresolved. I believe that the facts presented in this book also demonstrate the validity and relevance of this question as one of the most profound and important facing science today.

On a more personal note, Einstein's words have come ever more increasingly to reflect my own attitude. The deeper I delve into the mysteries of science, the more awe I experience at the magnificence of design I see reflected everywhere in the laws of the universe. Scientific data do not contradict the existence of God. On the contrary, for me, they overwhelmingly support it. With respect to the attitude of the scientific establishment, it appears to me that we are experiencing something similar to what occurred during Galileo's time, but in reverse. He lived during a period when belief in God was hindering science because it had become a dogma that dictated which phenomena should be studied and how the data should be interpreted. Today, we are living in a period where non- belief in God is hindering science in the same way. Like the papal doctrine of Galileo's time, the doctrine of materialism is a dogma that dictates which questions are regarded as scientific and how the data are to be interpreted. It is not based on scientific evidence, but is a creed to which all scientists hoping to maintain credibility in the scientific community must nevertheless pledge allegiance. This doctrine is clearly standing in the way of progress and I am challenging the scientific community to address it.

That materialist reductionism can continue to dominate so many fields of science, despite overwhelming evidence of its inadequacy, is nothing short of astounding. The perpetuation of this fiction parallels the children's fable concerning the "Emperor's New Clothes." Somehow a king allows himself to be convinced that the weavers to whom he has paid extravagant sums of money to weave new clothes, are weaving from a beautiful thread that is invisible to stupid people. They convince him that this is how he will be able to judge the intelligence of his subjects. Not wanting to appear stupid himself the king pretends to be able to see the clothes. His courtiers do not want to admit that they are stupid or unworthy of their favored positions, so they too profess admiration for the clothes, never openly questioning their invisibility. So, the king decides to show off his new clothes in a parade. The people on the street who see the naked emperor also do not want to appear stupid since everyone else appears to be seeing him clothed. So they, too, pretend to admire the new outfit. Suddenly a child says, "But he is naked," and the meaning of what they had been observing, becomes evident. The people are still seeing exactly what they saw before, but suddenly they are interpreting it differently. The clothes were not invisible they were simply not there. The premise of "no God" has assumed the properties of the emperor's new clothes by defining any scientist who seriously questions it, as ignorant and unscientific. No one wants to appear stupid, so most scientists (but not all) who do question it, and their number is growing, do so in silence. Science has become the new religion and anyone belonging to the biomedical sect of this church, who does not adhere to its doctrine of materialist reductionism, risks excommunication from the community.

In the name of science, it is time to educate the public about the facts and to put an end to this folly. It makes more sense to say that anyone familiar with the data I have presented, who doesn't at some point at least wonder about the existence of a Divine Creator, is not only lacking in objectivity, but is exhibiting psychological defense mechanisms bordering on denial. Faced with the evidence, seriously contemplating God's existence is not indicative of "magical thinking," ignorance, or lack of scientific rigor, but one of the most logical options available. Refusal to acknowledge this topic as a legitimate domain for scientific investigation is itself a sign of subjective bias. The complete silence on this issue currently reigning in the scientific community is impeding the progress of science, especially in the area of biomedical research. If the current hypothesis of the non-existence / irrelevance of God for science is correct, its proponents should welcome the challenge to finally put this speculation to rest. If the current premise is incorrect, however, then the existence of a Divine Mind or Cosmic Consciousness could be a tremendous source of healing, power and creativity, whose discovery would greatly benefit humanity. We owe it to ourselves and to the future of science to begin this exciting journey of discovery.

REFERENCES

8.1)   Planck M. In F. Ungar (ed.) *Practical Wisdom; A treasury of Aphorisms and reflections from the German.* Fredrick Ungar Publishing Co., 1977, New York, N.Y.

CPSIA information can be obtained
at www.ICGtesting.com
Printed in the USA
BVHW081041181119
564148BV00011B/260/P